The Book Publishing Company — Summertown, Tennessee 38483
UK and Eire Distribution, Kona Publications, London E.C. 1

*Inspired by the Friendly Stranger who helps us all get out.*

Publisher: The Publisher (Paul Mandelstein)
Authors:

   Stringbean (Mark Long) G5DEM, EI2VBW, WA4LXC
   White Lightnin' (Albert Houston) KW4A
   Minnesota Mumbler (Jeffrey Keating) WB4KDH

Editor: Cookie Monster (Matthew McClure)

Typographer: Neon Leon (Leon Fainbuch)

Production Manager: Daddy Longlegs (David Long)

Artists:

   Hush Puppy (Gregory Lowry)
   Mister Fixit (Peter Hoyt)
   Scribbler (Arthur Saarinen)
   Wooley Booley (Steve Solomon)
   Beast (James Hartman)
   Designer Genie (Jeanne Purviance)

Production:

   Tortoise (Tortesa Livick)
   Medium Apple (Jimma Egan)
   Hot Wire (Albert Livick)
   Massachusetts Mayflower (Nancy Holzapfel)
   Harmony (Hannah Scott)
   Dreamie Jeannie (Jeannie Egan)

      Thanks to all the other Big Dummies
      who helped make this book possible.

ISBN #0-913990-35-3
Copyright ©1981 The Book Publishing Company
1st Edition August 1981

Printed in Great Britain by
Richard Clay (The Chaucer Press) Ltd,
Bungay, Suffolk

Any reproduction of this book without the express permission of the Publisher by duplication processes, printed form, photographic or electronic or magnetic storage and/or retrieval systems is expressly forbidden. This book's first page is a watermarked end paper. Any copy without this watermark is a pirate edition which is in violation of the Universal Copyright Act of 1956.

The original *Big Dummy's Guide to C.B. Radio* has received international acclaim as *the handbook* on CB Radio. Due to popular demand, this new edition was expressly written to keep up with the latest developments in today's CB world.

# TABLE OF CONTENTS

**INTRODUCTION** .................................... 10

**CHAPTER 1 – GETTING ON THE AIR** ................... 13

    Modulating – Getting a Handle – Breaking the Channel –
Radio Check –S-Meter Reports – Frequencies –
Channels – Ten Calls – Squelching the Noise –
Skipland – Getting a License – Rules and Regulations

**CHAPTER 2 – BUYING A NEW RIG** ..................... 28

    FM vs AM/SSB Radios – 27 Megs vs 934 MegaHertz Rigs –
Knobs and Dials – Making Something out of the Advertising –
Specifications – Power requirements –
Mobile or Base? – Warranties

**CHAPTER 3 – MOBILES** ............................... 40

    Installing Your CB – Negative Earth Installation –
Positive Earth Installation – Anti-Theft Installations –
Mobile Antennas – Whips – Twin Antennas –
The Non Sus Antenna – UHF Mobile Antennas – Mounts –
Quickie Installations – Quickie Mounts – Hooking Up the
Coax – Connecting the Coax to the Antenna – SWR;
Is Your Antenna Working Right? – Checking SWR –
Adjusting Your Antenna

**CHAPTER 4 – BASE STATIONS** ......................... 54

    Installing a Base Station – Using a Rig as a Base –
Home Twenty Aerials – Gain – Effective Radiated Power –
Ground, Space and Sky Waves – Polarisation –
¼ Wave Vertical Antenna – ⅝ Wave Vertical Antenna –
Beams – Vertical Beams – Stacked Beams –
Switchable Horizontal/Vertical Beams – Quads –Coax –
Lightning – Towers and Masts – Rotators

**CHAPTER 5 – JOURNEY TO THE BOTTOM OF
                   YOUR RIG** ............................. 71

**CHAPTER 6 – FM, AM and SSB** ........................ 82

    Frequency Modulation – Deviation – The Capture Effect –
Kicking Out the Noise – Amplitude Modulation –
Overmodulation – AM Noise Elimination –
AM Receiving – Single Side Band

**CHAPTER 7 – DX GUIDE FOR CBers** .................... 90

Sunspots – The Sunspot Cycle – Seasonal Skip Cycle –
Daily Skip Cycle – Solar Storms and Radio Blackouts –
Fading – Long Path – North/South Skip –
Sporadic E and Other Exotic Skip – Burners –
The Eleven Metre Band – QSL Cards –
International 40 Channel Band Plan – SSB Lingo –
Q-Signals – RST Reports – Phonetic Alphabets –
Amateur Radio – Morse Code

**CHAPTER 8 – GIZMOS** ................................... 105

Power Mikes and Speech Compressors – Short Wave
Receivers – Scanners – Walkie-Talkies –
Power and Modulation Meters – FM and CB Receivers –
External Speakers – Antenna Switches – Antenna Matchers –
VFO – Frequency Counters – Receive Preamps or Boosters –
Bilateral Amplifiers – Radar Detectors – PA Horn –
Power Reducers or Attenuators – TVI Filter –
Dummy Loads – Tone Squelch – Phaser Lasers, Gooney Birds,
Roger Bleeps and Pings

**CHANNEL 9 – EMERGENCY PROCEDURES** .............. 116

**CHAPTER 10 – FIXING YOUR RIG OR WHAT
WENT WRONG** ...................... 118

Troubleshooting Guide – Repairing Microphone and
Antenna Connections – Coax splices –
Soldering Coax Connectors – Coax Splices –
Microphone Connections – Using a Continuity Tester –
Noise: how to get rid of it

**CHAPTER 11 – DO-IT-YOURSELF ANTENNAS** ............ 133

¼ Wave Ground Plane Antenna – Using a Mobile Antenna
for a Base – Building a Gain Vertical –
The Long John Antenna; Build a 10 dB Gain Beam

**CHAPTER 12 – CB IN EUROPE** ........................ 144

**CHAPTER 13 – HOW FAR CAN MY CB RADIO TALK?** .... 146

**CHAPTER 14 – CHANNEL JIVE (CBers Lingo)** ............. 148

**TIME AND METRIC CONVERSION CHARTS** ............. 155

**TEN CALLS** ............................................. 156

**BRITISH AND AMERICAN CB FREQUENCIES** ........... 157

**INDEX** ................................................. 158

More than 61 countries around the globe now have some form of CB radio service. And, after much discussion, both the British and Irish governments have decided to implement a CB radio service on 27 MegaHertz FM as their answer to the hundreds of thousands of underground CB radio stations already operating in their countries. For the first time in the UK and Eire, individuals can legally have their own personal communication service. Your radio conversations are free, and you don't have to know a lot of fancy radio theory to get on the air. CB is fast becoming the people's radio service and all kinds of folks are ready to dive in!

CB communication is not only fun; it's also smart. On the motorways, back home or in the office, CB is a useful tool for almost everyone. Lorry drivers and motorists, and the motorway patrol, use CB to keep things together on the road. It gives them an extra sense that stretches them on out miles ahead on the motorways and highways, so that they can perceive changing road conditions, weather, accidents and other driving hazards. Truck drivers use CB as a means of staying intelligent behind the fast paced, hammerdown lifelines of Europe.

CB radio lets you talk to your home from your mobile via the radio airways. It's convenient for saving time at stops, sending instructions, and relaying important phone calls and messages, as well as for helping you stay connected to your family. CB radio also offers a means of staying connected with your business. For many people, the added communication means added gain, too. CB communications add an extra means of being in the right place at the right time.

In an emergency, CB radio may be the only means of getting help. There are emergency groups (HELP, REACT, THAMES, HARP, etc.) that monitor the radio and can offer help and assistance. These folks also help in times of natural disaster. Many more police will be getting CBs too, making them more readily available to motorists when needed.

There are many CB clubs throughout Britain and Eire. Breakers Clubs like the UBA and the NCBCI have been instrumental in getting CB recognised as an excellent radio service for the public. Legalisation has come about because of dedicated efforts by groups such as these. They offer you the opportunity to get to know your fellow breakers off the air as well as on it. CB clubs have also joined together to offer assistance to folks in need of help, often giving aid to the elderly and the disabled. You can check around and find out where the local CB club meets in your area. We discovered that there are a lot of good folks out there to meet!

## CHAPTER ONE
### Getting on the Air
### or
### Basic Modulating

A CB radio is technically called a transceiver, which is a combination of a transmitter and receiver. Your car radio is a receiver that you can listen to—but you can't talk back to the BBC. CB radio can be more fun than a MW radio, because you get to communicate with the folks that you hear. Throughout the book we use CB, CB radio, rig, transmitter, hunk of junk, chicken box, squawk box, and the like, interchangeably with the words "Citizens Band radio."

### Modulating

Well, you can't actually hear the radio waves themselves. Your voice is hooked onto the radio waves by a process called modulation. Modulating also means talking on your CB. We use modulate, modjitate and ratchet jawing to mean the same thing. When you first get a rig you could give a listen for a while to get a feel for what's going on. If you want to jump right in there, though, don't worry—the other folks on the channel will let you know how you're doing. CB is a down-home mode of communication and folks will love modulating with you. If you don't understand all the lingo, check out our glossary on *Channel Jive*.

## Getting a Handle

Your handle is your CB alternate personality code name. You might only get to know the folks that you meet on the CB by their handles. One way to get a handle is to have your friends think one up for you, or you might just think of one on your own. You may go through a few, but eventually one will stick.

## Breaking the Channel

There are forty channels for use, and each one may have as many as hundreds of users in certain areas. With so many folks using CB, you need to make sure that the channel is not in use before bulldozing your way in. The most common way to do that is to "break the channel." Always listen to see if someone is talking. Then you can say, "Break one-four," (if you're on channel 14), and usually if the channel is free, someone will come back and say, "Go, breaker," or you might be asked to "Stand by," or "Hold on." If you don't hear anything, you can assume that the channel is clear and make your call. Other things you might hear for breaking the channel are "break, break," "Breakity-break," or "Breaker broke break," etc. Asking for and giving a break keeps the channels from drifting into total madness.

Turn that radio on. Take a spin around the dial. Listen to some of the people talking. Keep in mind that you are entering another dimension, and it would be wise to learn the ways of the people that you will be talking to. A few minutes of turning the channel selector should inform you on which channels are in local use.

If a channel is already in use, and you would like to join the conversation, wait for one of the stations to finish a transmission, and then quickly press your mike and say, "Station on the side!" One of the breakers on the channel might let you in by saying, "Station on the side, go ahead."

## Radio Check

One of the first things you'll want to do with your rig is to get a "radio check." This is when you call out on the channel for the purpose of finding out how well your radio is performing. A typical radio check might go like this:

> **Break for a radio check.**
>
> Go ahead, radio check.
>
> **10-4. You got the Big Dummy. Who we got there?**
>
> You got the Fox. You're comin' in good here—putting about nine pounds on my meter. And good modulation, c'mon.
>
> **10-4, good buddy. We appreciate the check. We'll back 'em on out. The Big Dummy, going breaker break!**

Folks can give you information in a radio check in two ways. They can listen to you and tell you how it sounds to them. If you are kind of weak they might say that they *"got you back in there."* A fair signal might get a *"definitely making the trip."* Strong signals can get you a *"wall-to-wall"* report, or *"you're bending my windows."* Another way to give a radio check is by using an S-meter (signal strength meter). Many radios have an S-meter on them (see illustration). Stations that you talk to will move your S-meter to varying degrees, depending on the strength of their signal and their location. "You're putting about nine pounds on my meter," means his S-meter is reading 9 when you are talking. For higher readings like 20 or 40 or so, folks sometimes say, *"You're pegging my meter,"* or *"You're in the red."* Ways to describe how you are getting out are as countless as the sands of the Ganges River.

## S-Meter Reports

Got you way back in there. (S-1)

You're a bit 10-1 on me, c'mon. (S-3)

Good copy— nice one. (S-6)

Wall to wall and tree top tall. (S-9)

Busting my speaker. (S-+20)

Blowing out my windows. (S+ & a broken meter)

## Frequencies

### The 27 MegaHertz Citizen's Band
### The 934 MegaHertz Citizen's Band

The Home Office has allocated these two bands of radio frequencies for use by breakers. When talking about a radio frequency on the 27 MHz CB band, we mean a regular wave with a frequency of 27 million cycles per second. That means 27 million waves, each 11 metres long, radiate from your antenna each second, traveling at the speed of light! This can be represented by waves like this:

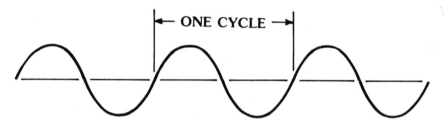

One cycle per second is also called a Hertz, named after the early radio pioneer, Heinrich Hertz, who first experimented with radio waves back in the late 1800's. One thousand cycles per second is also called a kiloHertz (kHz), and one million, a MegaHertz (MHz). Twenty-seven million cycles per second = 27 MHz.

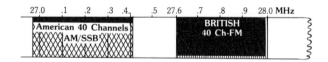

## Channels

There are forty channels for use in the 27 MHz FM band. These new British channels may have the same numbers as the older US AM CBs, but they are actually on different frequencies. A few of the new channels are set aside for specific uses by British CBers. These include an emergency channel, a breaking channel, and a traveler's channel.

***Emergency Channel*** – Channel 9 is used only for communications involving an immediate or potential emergency situation. It is recommended that you keep the channels on either side of the emergency channel clear, to prevent causing interference to the emergency channel.

***Breaking Channel*** – Each town or local community has one or more channels where breakers can listen or call for their friends. When using a breaking channel it is important to move to another channel as soon as possible, so that the breaking channel remains free for others to use. You can ask around and find out which channel is used for breaking in your community.

**Break, one-four**

Go Breaker!

**Hey, we thank you for the break. How about it Big Dummy? Are you on this Smokeytown channel, get back?**

Four fer sure! Who do we got on that end? C'mon.

**You would have the Lucky Lady here. Let's clear this breaking channel and kick it up to two-three for a ratchet, c'mon.**

Roger Dee. We gone to 23.

***Traveler's Channel*** – Channel 19 is the traveler's channel—it's used by people on the road to communicate about what's happening. The truckers keep in touch as to the traffic and highway conditions, accidents or hazards. They also use it for directions and as sort of a Yellow Pages to find out a good place to eat or when to park their lorry for the night.

Also truckers use Channel 19 to keep track of Smokey the Bear. It is very important to leave Channel 19 for the truckers and other mobiles that are traveling on the motorways and highways. With 40 channels available for everyone's use, it is only fair to give the travelers a break and keep 19 clear for their use, 10-4?

*Break one-nine for a copy!*

Go ahead breaker, we got a copy.

**10-4 good buddy, you got the Daredevil on this end...What's yer handle, mate?**

You got the One-Eyed Jack. What kind of wheels are you pushing c'mon?

**We're pushing big wheels, mate, and we got the hammer down for Spaghetti Town, is that a four?**

Aye, that's a four. We be southbound for the Smokeytown. You be careful round that Spaghetti Junction. There was a Jam Sandwich collecting green shield stamps when we passed through, fetch it back!

**That's a four, Jack. Appreciate the info. It's looking clean and green back towards the Smoke. So we'll give you the high numbers. Keep it clean and don't be seen. May the blue light never shine upon you. You got the Daredevil northbound and down!**

Four fer sure, Daredevil. 10-10 till we do it again. We up, we down, we gone, bye bye!

**The Jam Sandwich**

## Ten Calls

*How about it Big Dummy? Are you 10-8, break?*
On channel!
*What's yer rough twenty now Big Dummy?*
Ah, we just passed by that King's Cross Station, c'mon.
*You were a bit 10-1 on me that time, so give me a nine on that, c'mon.*
I said that I just passed the King's Cross Station, c'mon.
*Aye, that's a four, good buddy. We'll be 10-10 and listening in.*

Those breakers were saying "10-4" and "10-8." They are using ten signals, which are agreed upon radio shorthand that folks use throughout the world. In some cases the 10 part of the number has been dropped as in saying "give me a nine" instead of asking for a 10-9. The calls go from 10-1 to 10-100. Here are some of the more commonly used ten-signals.

| | | | |
|---|---|---|---|
| 10-1 | Weak copy | 10-13 | Weather and road conditions |
| 10-2 | Loud and clear | | |
| 10-3 | Stand by | 10-20 | Location, also rough location |
| 10-4 | Yes, Okay, Roger | | |
| 10-5 | Relay, Repeat information for another station | 10-27 | Moving to Channel ___ |
| | | 10-30 | Illegal use of radio |
| 10-6 | Busy at present | 10-33 | Emergency traffic on the channel |
| 10-7 | Out of service, Off the air | | |
| | | 10-34 | Request for assistance |
| 10-8 | In service, On the air | 10-36 | Correct time |
| | | 10-46 | Assist motorist |
| 10-9 | Repeat | 10-77 | Negative contact |
| 10-10 | Standing by | 10-100 | WC stop |

*The 10-36 is...*

If you hear a 10-33 on any channel, that means there is emergency traffic happening on that channel. Try to help out. This may mean just being quiet. Also there is another expression, "Code Three," which is used when talking about emergency vehicles, using lights and siren. "There's an ambulance southbound running code three!!"

## Squelching The Noise

Between conversations on your radio you will hear a lot of noise and static. The squelch knob on your radio can be set to turn off this noise and yet let you pick up any breakers.

Here's how to set it:

While listening to the background noise coming out of the radio's speaker, adjust the squelch knob clockwise slowly until the background noise just disappears. Leave the control at this setting. The receiver will remain quiet now until someone transmits a signal on your channel. That signal will "break" the squelch and be heard by you. Remember, it is important not to turn the squelch control too far beyond the noise cut-off point, or some of the weaker stations on the channel will not break through it.

Unlike AM, on FM you won't miss anything by setting your squelch this way. On FM if a station isn't strong enough to break the squelch, it won't be strong enough for you to understand clearly anyway. So don't feel that you have to listen to that old receiver hiss in order not to miss anything, because you don't. If you do, after a while the noise will drive you crackers!

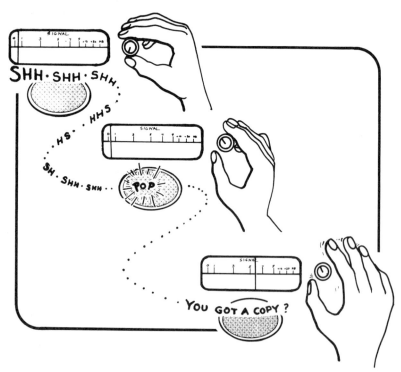

## *Skipland*

On some days, you might hear people talking from far away places not within the usual range of your radio. You can receive these long-distance stations because the radio waves are bouncing off the ionosphere, the high atmosphere around the earth. The sun causes layers in the ionosphere to become electrically charged and act like a mirror for radio. At certain times, the radio waves can be reflected back to the earth, skipping great distances with leaps and bounds. That's why we call it the "skip."

Stations talking skip use unusual names and numbers. They are also probably using gigantic antennas and high-powered transmitters. If you try to talk back to them from your weak little mobile, they may not hear you. Don't be alarmed. Legally, you're not supposed to talk that far anyway. On your legal FM radio AM skip talkers will not be heard at all, and SSB stations will sound all garbled. It may be hard to find a clear channel in times of heavy skip. Just wait a while, and the skip will roll out just like it came in. Usually there is not as much skip at night.

## Paying Your Dues To the Home Office

### Getting a License

The Home Office, Radio Regulatory Division, is in charge of administering the airwaves. Portions of the 27 MHz and 934 MHz bands have been allocated for the Citizen's Band Radio Service.

The operation, importation and installation of unapproved CB radio equipment is illegal under the Wireless Telegraphy Act of 1949 and 1968. Violators of this act are liable for possible jail sentences, fines and confiscation of illegal equipment—Mercy Sakes!

The government issues licenses. Each owner is given a call sign after filling out an application and paying a license fee. The Post Office Radio Interference Service (a division of the GPO), has set certain technical standards to ensure both the effectiveness of CB and the minimum amount of interference to other radio services. All stations are required by law to use only **type approved** radios that are manufactured for 27 or 934 MHz FM. All previously imported AM/SSB radios remain illegal for use in the UK. It is the responsibility of the owner to ensure that only **type approved** radios are used. All **type approved** equipment will have a 27/81 circle stamped or engraved on the radio's front panel.

The new license makes you responsible for whatever happens on your rig. Not everyone in your family will need to get a license— one license covers it all. If you have a business, one license can cover your employees as well. No exam is required.

The Home Office wants to keep Citizen's Band a good reliable means of *local* communications; it is their position that if people want to communicate over long distances and make international contacts, they should become licensed radio amateurs by taking the amateur radio exam. DX contacts on CB are considered an illegal offense.

All British 27 MHz CB antennas are limited to a single element rod or wire antenna with a length of 1.5 metres (59″) or less. This effectively limits the range of local as well as DX communications. All other antennas used by CBers on the continent and in the US are illegal for use on the British CB channels. Antenna height is also restricted to below 7 metres (23 feet) unless a power reducer or attenuator is used to limit the transmitted signal to less than one-tenth of its unattenuated amount. This would take your 4 watt signal down to *toy power* levels of less than ½ watt. All type approved CB radios are required to have a switch or other means of easily accomplishing this.

The Home Office decided to put its 27 MHz CB service on frequencies located above the regular American AM/SSB channels. They felt that by doing this, a more reliable service would be possible, eliminating much of the interference caused by skip and the illegal local operation of AM/SSB equipment. It is also the only 27 MHz CB service in the world on these frequencies!

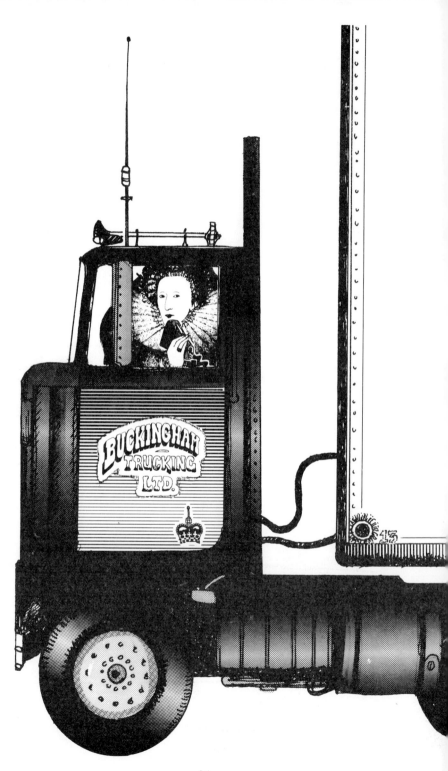

## Get A Load of These Rules & Regulations

A license from the Home Office, Radio Regulatory Division, is required for any CB radio equipment capable of transmitting. This license will need to be renewed periodically.

All CBers are required by law to purchase and use only CB transmitting equipment that has been **type approved** for use within the UK.

You are expected to say your assigned station identification numbers at the beginning and end of each transmission. Handles are not a bona fide substitute for your call numbers.

Never transmit the word May-Day or any other international distress signal unless there is a confirmed grave and imminent danger to life or property.

You should not in any way intentionally interfere with the communications of other stations.

Keep transmissions brief and to the point.

The use of CB radio for anti-social purposes is prohibited.

Do not use obscene or profane language on the air.

The use of CB for broadcasting or the reproduction of music is prohibited.

If we remember how many fellow CBers there are, and that we're all in this together, everything will be all right! 10-4!

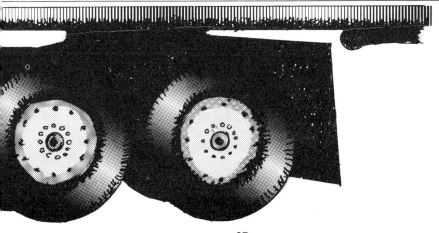

# CHAPTER TWO

## Buying a New Rig

CB radios can be constructed in many different ways and come with lots of "extras." These extras, most of which are used for listening, make the difference in the price of most radios. The many combinations of these features makes for a lot of different CB radios on the market. Now it doesn't matter if you are driving a Volkswagen or a Porsche, you can still get around. And with a moderate expense you can get into the CB action. In fact, the reason that we call these features "extras" is because they are extra and not necessary to get out well.

Legally you aren't supposed to improve the transmitting quality by boosting the power. So manufacturers try to improve the circuitry in other ways, and also add things that let you pull in weak stations or cut down on noise and interference—such as a noise limiter. Other extras and added features do help, but are not critical. Don't get snowed into thinking you need a crystal lattice filter or range expander in order to get out, because it isn't true.

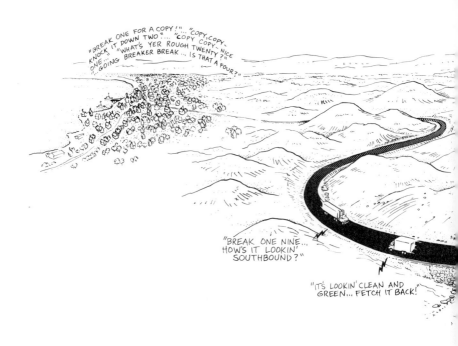

A *transmitter* is the generator of your radio signal, but it's the antenna that both catapults your signal outward on transmit, and captures the signals on receive. So keep in mind that you are going to need a ***good antenna*** and coaxial cable. This is one of the most important factors in how well your radio is going to get out.

The power that your radio will put out affects its performance and its price. The power used by a CB is expressed in two ways: as power input and power output. The power input is the amount of power used by the transmitter to produce what goes out. The power output is the amount of power which actually gets out of the radio. Power is measured in watts. The Home Office has set a limit of 4 watts output power which is what most new rigs deliver. So that's the best you can legally do. The amount of power you put out affects your range of communication. Four watts output power to the antenna can provide relatively good communication in a radius of 10 to 20 miles or more. Limiting communications to low power was designed to improve quality of local CB communication by cutting down interference from the traffic of neighbouring communities. It also allows there to be many folks on the air at once.

Now maybe your aren't too concerned about getting out the maximum distance possible. There are cheaper rigs with fewer watts. The most common kind is the walkie-talkie which puts out less than one watt. Walkie-talkies' lower power combines with the smaller antenna size and lower antenna height to keep them from getting out as well as other rigs. A stronger rig will help talk around hills and buildings, and will cut through the interference better, so it never hurts to have more watts. It is usually worth the added expense to have the most power out that you can get.

## *FM vs AM/SSB Radios*

When talking into your CB mike, you are adding your voice to a radio wave. FM (Frequency Modulation) and AM (Amplitude Modulation) are two of the methods used to accomplish this. While either one is technically adequate for the job, only FM CBs are legally allowed in the UK. This is definitely a consideration when buying a new rig!

The UK FM channels are not the same as those used in the United States and on the continent. Within Britain, there are several hundred thousand breakers operating illegal AM CBs on the internationally established AM channels. Many of these UK breakers are even talking skip with CBers in other parts of the world. The Home Office says that this causes interference to other radio services, including hospital paging systems and radio controlled airplanes. That is why the government decided to move the FM CB service to another section of the 27 MegaHertz Band.

SSB (Single Side Band) is a stronger, more compact form of AM and it is very popular among those CBers who like to talk skip. Even locally, a SSB signal has 1½ to 2 times the range of an AM or FM signal of equivalent power. For every AM channel, there are two SSB channels: an Upper Side Band (USB) and a Lower Side Band (LSB). Like AM, SSB is illegal in the UK.

## *27 MegaHertz Rigs vs 934 MegaHertz Rigs*

The Home Office has licensed CB communications on two portions of the radio spectrum: 27 MHz and 934 MHz. Twenty-seven MHz is considered a High Frequency (HF) Band, and 934 MHz, an Ultra High Frequency (UHF) Band.

**27 MHz.** Within the UK, there are forty channels to use, with a maximum of 4 watts output power to the transmitter section. Because the 27 MHz FM equipment is a lot cheaper than the 934 MHz gear, most of Britain's new CBers will be located here. This means that it will be a lot easier to get a comeback when using the breaking, road or emergency channels on 27 MHz, than it would be on 934 MHz. Skip and skip interference does occur on 27 MHz. AM and SSB signals are not understandable on your FM receiver, while FM skip is!

**934 MHz.** On UHF frequencies, all skip interference is eliminated. Radio waves on 934 MHz are too short to skip off the ionosphere (they just kind of slip on through). 934 MHz radios have twenty channels for use, and eight watts of input power is the maximum allowed. The higher power helps to offset some of the disadvantages of UHF. Unlike HF radio waves, which can travel along the ground, around buildings and over hills, UHF communications are strictly limited to line of sight. All UHF antennas must be located at high and clear elevations for good results.

UHF radios are the product of new and sophisticated communications technology and their cost is several times that of the more common 27 MHz radio. The higher cost of this equipment means less crowded channels and is suited to certain business applications, where a more private means of communications is desired. UHF radios operate on frequencies similar to those used by microwave ovens. There is still some question as to the long term health effects from the user's exposure to microwave radiation. The National Radiological Protection Board advises against placing your head closer than 10 centimetres to a transmittting 934 MHz antenna or damage to your eyes may result. All 934 MHz walkie-talkies are limited to 3 watts or less as a precautionary measure.

## *Knobs and Dials*

Let's take a gander at a new CB radio. It has all kinds of fancy knobs and dials and switches on it, and you might be wondering what they are all for. Here is a list of common ones and what they do. We chose to show the rig below, because it has just about every feature you'd find on any CB radio. The British type approved rigs do not have the AM/SSB functions.

# RECEIVING

None of these affect how well you will get out.

**Volume, Off/On** - turns your set on. It adjusts the loudness of what you are listening to.

**Squelch** - filters out background noise, static, and weak stations. There's a threshold level, and if you turn to just past the threshold you'll cut out most of the noise but still get most of the strong stations on the channel. The further you go past that threshold, the stronger a station has to be in order for you to receive him, and you may miss out on some of the calls.

**Channel Selector** - selects which of the possible 40 channels you are transceiving on. Push button control and lighted readout have become popular. Some radios have push button control scanners that will keep moving from channel to channel with the button held down. For convenience, some companies are putting the channel selector in the mike, a repairman's nightmare.

**Automatic Noise Limiter and Noise Blanker (ANL & NB)** - these switches cut down on static and ignition and motor noise. They help filter out noise interference to keep it from getting into your receiver and making it harder for you to hear.

**Delta Tune** - is a control provided on some transceivers which make it possible for you to tune in stations that are a little off the centre of a channel. Try different settings of this while listening to the station you want to hear.

**RF Gain Knob** - cuts down the RF (radio frequency) volume in your receiver amplifier. It can be adjusted to cut down noise on nearby stations. It has a knob which gives you lots of possible settings. Run with the RF gain up all the way unless you get close to a station and it starts sounding so loud that it distorts your receiver.

**Distant-Local** - is an attenuator that in the Local position prevents local up-close CB radio traffic from overdriving an AM receiver. The Distant position is the normal one which gives you the full range and sensitivity of your receiver.

**Meter** - most meters have two scales—one for transmitting and one for receiving. One lets you know how strong a signal you are putting out and the other lets you know how strong a signal you are picking up. The first one is a *relative power out meter*. The second is a *signal strength meter* (S-meter).

**Mode Switch** - some radios are manufactured to allow the operator to switch between AM, SSB or FM modes of operation. These units transmit and receive the selected mode.

**Clarifier** - this control is used on SSB CBs only. Unlike AM or FM receivers, which are locked onto the channel frequency, the SSB receiver must be fine tuned for best reception. This control normally does not change the transmitted frequency.

**LED Digital Readout -** this display gives the channel number a lighted digital readout.

**Hi/Low Tone Switch -** this control acts like a sound filter on your receiver. It is similar to a bass/treble control on a stereo, and it allows you to select the sound that you like best.

**Hi/Low Channels -** many of the AM and SSB radios manufactured in Asia have this switch on them. In the low position the radio operates on the regular 40 AM channels. When placed in the high channel position, the radio will operate on 40 more channels that are located above the regular AM 40 channels. In many cases the LED Digital Channel Readout will then count from channel 41 to channel 80.

**Channel 9 Switch -** this control allows you to switch instantly from any channel to the channel 9 emergency frequency. When this switch is engaged, the regular channel selector is inoperable.

**Channel 9 priority -** when this is switched on, your receiver automatically listens every second or two on channel 9, and if anyone is transmitting there, the receiver stays on 9 until they're done transmitting. Usually the squelch must be set properly for this to work. This feature is usually only used by channel 9 emergency monitoring operators, and it must be switched off for normal operation.

# SENDING

**Microphone -** the push-to-talk button is on the mike. You have to push it to transmit and let go of it to receive. You can't hear anybody if you're holding that button in! But they will hear you if you talk clearly and at an even tone close to the front of the mike.

**Mike Gain -** increases the strength of your modulation. Ask somebody for a radio check to tell you at what point your signal starts to distort or break up as you are increasing the volume. That will let you know where the best place is to run that thing.

**P.A. -** allows you to turn your CB into a public address system. There is a jack in the back where you can plug in a cord that runs to a loudspeaker. You can even put that loudspeaker under the bonnet of your mobile if you want.

**LED Indicator Lights -** these small coloured lights are used to indicate the proper functioning of the transmit, receive or modulation.

**Antenna Warning Light** - if your antenna stops functioning properly, this light will come on when you transmit, to let you know that there's something wrong with the antenna.

**Transmit Attenuator** - when switched on, this limits your transmit power to 1/10 of what you would normally get.

## *Making Something out of the Advertising*

Now suppose you are interested in getting a CB radio but don't know much about them. You could be strolling through the neighbourhood shopping centre and be surprised by a big display of CBs. You try to stretch your imagination and comprehend all the advertising boasting of the quality of their inner workings, but it leaves you at a loss. Well, don't let that bog you down. Here are some common features you may run into:

***Dual Conversion IF*** - a little fancier receive circuit that gives some added clarity of reception over single conversion.

***Filters*** - ceramic, crystal lattice and mechanical; these are different ways of filtering your receiver in order to prevent *bleeding over* of conversations on channels next to the one that you are using.

***Automatic modulation compressor circuit (and limiter)*** - this gizmo maintains a high level of modulation over a wide range of voice loudness. It also prevents over-modulation. A requirement on all new radios.

***Type Approved*** - only radios marked *Type Approval* by the Home Office will be legal for use within the UK. This is indicated by an authorised stamp that is engraved onto the radio.

***Specifications*** - these are usually talked about with a lot of fancy figures thrown in that add to the confusion. When buying a new rig, you should try to get the best ratings you can.

UK Type Approved Rigs should have ample *selectivity* for most applications and have equally good *sensitivity*. *Selectivity* is the receiver's ability to differentiate between an adjacent channel signal and the desired one, so folks aren't bleeding over on your channel. *Adjacent Channel Rejection* also has to do with this. *Sensitivity* is a measurement of how well your radio can hear. Good sensitivity can really make a difference for a base station and give you a longer range of reception. For mobile stations it is less critical since the ignition and static noise of the vehicle will sometimes drown out the weak stations that would have been picked up by a sensitive radio.

Here's a typical set of specs for a good quality rig and what they mean.

## THUNDERBOLT 3000 SPECIFICATIONS

Sensitivity – 0.5uV for 10 dB S+N/N; Selectivity – 6 kHz at -6dB; Adjacent Channel Rejection – 50 dB at ±10 kHz; Frequency Tolerance – ±.003%; Image Rejection – 60 dB; Squelch Sensitivity – .2 uV; Audio Power Output – 2 watts at 10% THD; FM Deviation – ±2.5 kHz; RF Power Output – 4 watts; AM Modulation – 90%; Spurious Output – -50 dB max.

**Sensitivity – 0.5 µV for 10 dB S+N/N**

Sensitivity gives you an idea how good it is at pulling out a weak signal. The Sensitivity, "µV" means microvolts—the lower the number, the better. The conditions. It's the "signal plus noise-to-noise ratio." The larger the number of dBs, the better.

**FM Sensitivity – 0.3 µV for 20 dB noise quieting**

The FM sensitivity rating tells you the amount of signal at the antenna input necessary to be 20 dB stronger than the received noise level. This is the minimum signal that would fully quiet the background noise and make for a readable copy. Again the sensitivity is measured in microvolts and, the lower the number here, the better. The industry standard.

**Selectivity – 6 kHz at -6 dB**

These two are related and they have to do with how clear your channel is gonna sound if someone is using the channel next to yours. A lower figure here is better. They're usually between 5 and 6. The usual conditions.

**Adjacent Channel Rejection – 50 dB at ±10 kHz**

Some manfacturers don't give selectivity the way it is here. They use the term Selectivity instead of Adjacent Channel Selection. A larger figure here is better, like 75 dB. The usual conditions.

### Frequency Tolerance - ±.003%

This gives an indication of how much the radio can drift off the channel, under normal operating conditions. On FM it is important to be as close as possible to the channel frequency for best operation. The lower this percentage, the better the stability of the radio will be.

### Image Rejection - 60 dBs

Unwanted signals from above and below the CB band can penetrate into a receiver's IF section, interfering with what you can hear on the channel. The larger this figure, the more interference free the radio's receiver will be from unwanted image frequencies. Dual conversion radios have much better image rejection than single conversion units.

### Squelch Sensitivity - .2 μV

This is the minimum received signal at the antenna terminals that can trip the squelch control. This figure is slightly lower than the sensitivity rating of the receiver. The lower this figure, the better.

### Audio Power Output - 2 watts at 10% THD

The CB has a little amplifier just like a stereo and this is how much power it has to drive the speaker. The more watts, the louder the it can sound at full volume. 2 or 3 watts is plenty. This is the CB standard for clean sound. THD means Total Harmonic Distortion.

### FM Deviation - ±2.5 kHz

This indicates the amount of frequency shift above and below the channel frequency that is generated by FM modulation. 2.5 kHz deviation is a Home Office standard.

### RF Power Output - 4 watts RF power is the UK maximum.

The basic power of your transmitter. RF power output is how much radio frequency energy power actually comes out of the coax connector of the radio when transmitting. The more the better.

### AM Modulation - 90% typical

This is how much of the available talk power the transmitter actually uses. 100% is ideal, but they can't cut it that close on an assembly line.

### Spurious Output - -50 dB max.

This is the kind of garble that makes a mess on other channels, TV as well as CB. It's usually not mentioned. This is the GPO minimum. The larger this figure the better.

***Power requirements*** – if you are getting a radio for your car, get one with 12-14 volts DC rating. For a base station you can get a radio that runs off 240 volts AC. Some companies make radios with dual power supplies for both mobile and base station operation. *Positive or Negative Earth* means it can be hooked up easily to either a positive or negative earth vehicle.

### Mobile or Base???

You can use a mobile radio as a base station by using a 12-volt DC converter that plugs into the wall. A converter (power supply) costs about £15-25. This will work just about as well as a more expensive base station radio would.

## WARRANTY

Every new radio should come with a warranty. The warranty guarantees the radio to be free from defects in materials and workmanship for a certain period of time, usually from three months to a year. There is a warranty registration card that you need to fill out and send back to the manufacturer within the first few days in order to register your radio. Also included is information on how to return the unit for repair under warranty. In many instances, this involves returning it to the equipment dealer that sold you the unit. Make sure that you understand how this works. Don't forget to mail your card in!

Now that you got past all that, you will want to try some of those fancy beauties on for size. Ask your dealer if you can try one. Grab that old microphone and ask some local CBer how you're getting out.

**Breaker 14 for the Big Dummy. Big Dummy, you still on the channel, c'mon.**

On Channel! I just managed to make it through these pages, mercy sakes!

**Roger Dee. How do you like that Badger's rig there, c'mon?**

Well it seems to be doing good, is that a four?

**That's a four, mate. You're coming thru wall to wall, c'mon.**

Roger Dee, Roger Dee. We're going to get one of these. So you can catch me later tonight. I'll be mucking about in my mobile, trying to fit this box and a twig, c'mon.

**Yeah, that's a rog. I'll give you the golden numbers then. Catch you later, we go breaker break!**

# CHAPTER THREE
## *Mobiles*

CB has gotten as big as it has because of its great possibilities for communications while you're rolling down the road. It's great for talking back to your Home 20 and finding out how it's going back there, as well as getting connected with the flow of things out on the motorways. Since a lot of folks are going to want to start out with a rig in their four wheeler, we'll get right into how to install one in your car and mount a good antenna on it, too. So let's have a go at it!

### *Installing Your CB*

There are several good things to remember when you are going to put a new rig in your mobile. The first thing that comes to mind is where to mount it. This often depends on the driver's preference. There are some practical things, though, to keep in mind. You'll want it to be within easy reach and clearly visible when you're driving: you don't want to end up in the ditch because you had to dive for a mike to answer some local breaker. It should be out of the way of the gear shift or emergency brake. Make sure the mike cord and other wires won't get tangled up in the steering wheel or the accelerator and brake pedals, and that it is in a place where it won't get kicked or sat on. Also, heat can damage a rig, so don't mount it right under your car heater. So it usually turns out that the rig is mounted somewhere under the dash close to the driver, or occasionally suspended from the roof near the driver.

Once you have figured out where you'd like to mount it, drill the holes for the mount, being careful not to hit anything behind the dashboard, and screw that bracket up there and bolt the CB onto it. Now is a good time to figure out how this rig is going to get its electricity. Most radios have two wires coming out of the back for the DC power: a red one and a black one. Usually, red is positive and black is negative. Most radios with two wires like this can be used for either positive or negative earth vehicles. (Check the instructions accompanying your radio to make sure.)

It's important to check to see whether your vehicle is positive or negative earth. Just lift the bonnet and check to see which battery terminal is attached to the frame of the vehicle. Most cars are negative earth (the frame being earth) but **make sure first!!** Hooking your CB up the wrong way can damage it—set the ole set a-smoking.

### Negative Earth Installation

Get some insulated wire like the kind sold at automobile parts stores for automotive wiring systems. Run it directly from the positive lead of the CB to the positive terminal of the car battery. Add an in-line fuse holder to this wire close to the battery to protect your car's electrical system. Make sure you tape all connections with good quality electrical tape. The negative wire can be bolted to the metal body of the car somewhere, because the negative terminal of the battery is also fastened to the metal. This earthing will supply the negative connection to your radio.

The other way to do it is to tap into the car fuse panel (see picture).

This makes it so you only have to run a wire from the fuse panel to the positive wire of the radio, a shorter distance than all the way to the battery. It's a good idea to have a separate fuse for your CB. Most CBs come with a fuse in their supply lead. Never use anything bigger than a 3-amp fuse for a transistorised radio. Having a fuse can possibly save you some future repair bills by offering added protection.

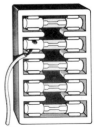
**Red Wire**

## Positive Earth Installation

Some big trucks and semis are *positive earth*. Positive earth installations require another method of hooking up your radio.

If you haven't got a rig yet, definitely check and see whether your vehicle is positive or negative earth. If it is positive earth, get a rig that can operate either way and it will save you a lot of trouble. Check the owner's manual to determine if your radio can be operated either positive or negative earth.

Most radios have two small wires, one red and one black. In this case, just hook the red wire to some bolt or screw on the body of the vehicle. Hook the black wire through an in-line fuse holder to the negative terminal of the battery, or to a hot terminal in the fuse box.

Some radios just have one small wire coming out of the back. This is the hot lead. It is usually positive. This type of radio can only be used in a negative earth vehicle, because the case of the radio provides the negative earth connection. Some radios have a switch or can be rewired inside for positive earth. Read the instructions for this kind of radio carefully. Always make sure to use the proper fuse if you have to experiment.

If you have a vehicle with only 6 volt DC, there are 6 to 12 volt converters that you can buy in order to feed your rig the right menu.

> Mercy Sakes! Don't turn on your rig unless it's connected to an antenna made especially for CB. No other antenna will do; not the AM radio one; no, not your TV antenna either; nope, only the real thing.

## Anti-theft Installations

Mobile CB radios just happen to be small, light, expensive and easy to rip off. Here are some suggestions which may help prevent this from happening to you. You don't want to lose your rig, for sure.

Your CB can be installed in such a way that you can easily remove it from under the dash and lock it in the boot or take it into your house. One way to do this is to use a *slide mount*. This is the kind of mount that allows you to slide your rig in and out easily.

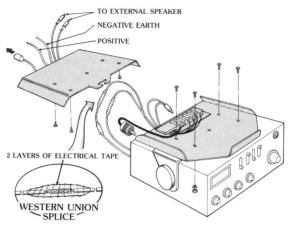

One section of the slide mount is attached to the mounting bracket of the radio, and the wires from the radio attach to a set of contacts on the slide mount. The other section is mounted in the vehicle and its contacts are wired into the vehicle's power and sometimes into the antenna. The two sections slide together and make contact. If you use one with the antenna connections on it, beware of the cheaper models that do not make good enough connection, resulting in a high SWR.

If you prefer to keep your radio in your mobile, you can get a *locking mount* that makes it so you need a key to take the rig off the mount. There are also easy-to-install car burglar alarm systems available at electronics stores. Some burglar alarm systems honk your horn or ring an alarm inside your car when the door or bonnet is opened, or if your radio is taken from its mount. When you leave it in your mobile, be sure to lock the doors.

## Mobile Antennas

The heart of CB communications is mobile-to-mobile and mobile-to-base operations. The twigs you have on your mobile definitely play an important part in how you get out.

One thing to keep in mind is that all antennas designed to operate on 27 MHz will work equally well on FM or AM. Antennas for the new British FM frequencies will need to be a little shorter in physical length than their American equivalents.

There are many different kinds of mounts and places you can put your antenna. You could stick that ear just about any place on your mobile, but some places have advantages over others.

The metal body of your vehicle is actually a part of your antenna. The location of the antenna in relation to the car body will affect the radiation pattern of your signal, which will be the strongest across the longest portion of the vehicle.

## Whips

By far the best physical length for a 27 MHz CB antenna which would still be practical for mobile use would be 2.5 metres (100″) long. Although we've seen these big twigs on the tops of some vehicles, they tend to mix it with the tree limbs and garage roofs, etc. So 2.5 metre whips are usually mounted on back bumpers. A 2.5

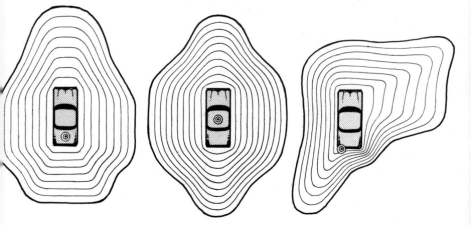

metre whip is equal to a quarter wavelength for 27 MHz. This whip, when mounted on the back bumper, uses the car body as part of the antenna and works best in the direction towards the front of the car. As a general rule concerning mobile antennas, the longer that they are, the better. And the higher on the vehicle you mount one, the better it will get out.

## Loading Coils

A loading coil is a coil of wire imbedded in a plastic cylinder. For *shortened* antennas, (any antenna less than 2.5 metres), the loading coil makes up the difference for the missing length. In the case of a 1.5 metre whip, the loading coil electrically replaces one metre of the total ¼ wavelength (2.5 m). The loading coil is connected to the *hot* centre conductor of the coax, and the steel whip or *stinger* is sticking out of the coil.

In many cases when a 2.5 metre whip is not practical, a shorter whip antenna is used. These short whips come in many varieties, but they can be divided into 3 main groups: base loaded steel whips, centre or *top* loaded steel whips and wire wound fibreglass whips.

### Base Loaded Steel Whips

Base loaded steel whips are generally very durable and inconspicuous. While they do not get out as well as a big twig or a *top* loaded antenna would, they do quite well for talking local, and have been used by CBers around the world for years. When mounting a base loaded antenna, keep in mind that the base coil radiates much of your signal, so it should be placed in a position clear of surrounding metal surfaces. On the roof, boot lip, or fender (cowl) mounting is the most common.

## Centre or Top Loaded Steel Whips

The centre or top loaded whip is the next best thing to a full-sized twig. They come in many shapes, styles and colours. The loading coil in the centre or near the top of the antenna is more electrically efficient than base loading for getting that signal into the air. Stainless steel is the best for long term durability. Some top loaded whips have an aluminium shaft that can be prone to bending. An impact spring at the bottom might be a good idea if your particular installation is likely to encounter low tree branches, etc. Then your antenna's survival might depend on it! Most loaded whips have a top section or *stinger* that can be adjusted in length for tuning the antenna.

## Fibreglass Whips

The wire wound fibreglass whip has become very popular lately. It comes in many colours and lengths. The most popular is the 1.2 metre (4 foot) whip. As a general rule, the longer the whip is, the better it will get out, whether or not it is called a ¼, ⅝, ¾ or full wave antenna. A full length ¼ wave fibreglass whip is 2.5 metres long. A base spring is recommended when using some fibreglass antennas. Fibreglass whips generally come tuned up from the factory and need little or no adjustment in their length. These antennas have a wire imbedded in the fibreglass that actually does the radiating of your CB signal.

This wire is sometimes wound spirally up the fibreglass rod forming a continuous loading coil. Of all the short antennas available, this one gets out the best. Some of the helically wound fibreglass whips look very much like a conventional MW car antenna, and are very popular.

## 1.5 Metre Whips

The Home Office has set a limit on the length of all 27 MHz antennas to 1.5 metres (59") or less. So this is the longest length that you can legally use in the UK—for base as well as mobile antennas! This short length makes the 1.5 metre whip a less efficient antenna when compared to the longer ones in use in other parts of the world.

## Twin Antennas

Twins are very popular on lorries, caravans and other vehicles where it is not possible to mount a single antenna on the roof. Two identical antennas are used with a special set of coaxial cables called

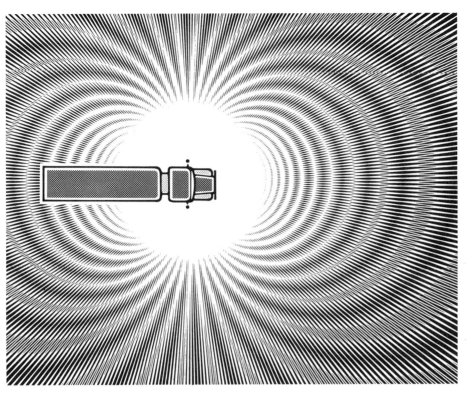

a *co-phase harness*. This is nothing more than two equal lengths of coax, both hooked to the same connector.

The two antennas combine together to provide a directional signal. If the antennas are at least 2.5 metres apart, such as on the two mirrors of a big lorry, they will get out best forwards and backwards. If they are *less than 2.5 metres apart*, they will get out *sideways* (something you wouldn't want when you are driving down the motorway). Remember that whenever you use twin antennas you end up sacrificing some directions in favour of others. It's not wise to use gutter mount twins on a car because you get out sideways. You would be better off using a base loaded antenna on the roof or boot of your car, or just a single gutter mount.

### *The Non Sus Antenna*

There are several disguised CB antennas that are made to look like regular car radio antennas. These can be quite handy in preventing radio thieves from knowing what you have in your vehicle. Some have a junction box which allows you to plug both your CB radio and your car's long and medium wave radio into the same antenna. Within the junction box, there are some tuned circuits which help to match the CB to the antenna. Usually these circuits are adjusted after installation, with the aid of a SWR meter. The trouble with this junction box is that you end up losing some of your power right there in the junction box. Also, if these circuits aren't set just right, you won't get out well at all, and could even damage your rig.

There are other kinds of disguised antennas that use no matcher and are not to be connected to the car radio. They perform much better because the antenna is pretuned at the factory for CB frequencies. All of your transmitter power goes directly to the antenna without going through any junction box or splitter. And best of all, they look just like the standard car antenna which they replace. In order to get your car radio back to receiving Auntie Beeb, you can get a MW antenna that adheres to the glass of your back window.

## UHF Mobile Antennas

On 934 MHz, the antenna's length is much shorter, usually less than ¼ metre long. The antenna must be mounted on the top of the vehicle, clear from any obstructions (i.e. luggage racks, car aerials, etc.), for good results.

Due to more exacting technical aspects, installation and maintenance of 934 MHz stations should be left to experienced radio technicians.

## Mounts

The best place to put a single antenna is smack dab in the centre of the roof. This is the highest place you can put it. This gets it up and out of the way of obstacles that could interfere with the radio waves coming off the antenna. The metal frame of the car is used as part of the antenna (called earth). Mounting the antenna in the centre gives a balanced and uniform radiation pattern which will allow you to get out well in all directions.

A mount is a device that hooks your antenna onto your vehicle. Mounts are usually made of metal and have a plastic part which electrically insulates the whip from the metal. There are many kinds of mounts—almost as many as there are antennas. The most popular ones are the boot or trunk lip mount and the rain gutter clip mount. The rain gutter mount is good for getting your antenna up high where it can get out. The boot lip mount can also be used for mounting to the bonnet. An important thing to remember with any mount is that it must have a good electrical connection between the mount and the car body.

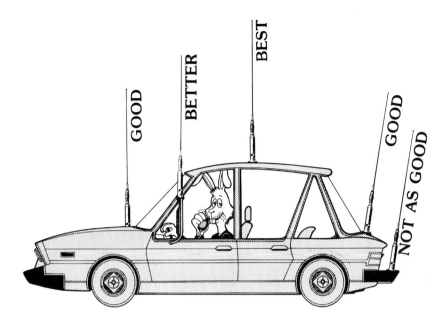

Some mounts have an adjustable set screw which is tightened until the pointed end of the screw pokes through the body paint, electrically bonding the mount to the metal of the vehicle. Some other types of mounts require boring a hole through the body of your vehicle. Now don't cringe, for heaven's sake. It isn't that bad. This is the strongest method of mounting a twig.

The mirror/luggage rack mount is a good mount to use if one fits your vehicle. Bumper mounts are useful for longer twigs on coupes or small cars.

The centre conductor of the coax connects to the whip—either directly, or through a loading coil. Neither the whip nor the centre conductor should touch other metal. If you think that the whip might touch the metal body when your vehicle is in motion, you can insulate the section of whip with some electrical tape or plastic tubing. Avoid installations where the antenna whip is up next to the vehicle's body for over a ½ metre of the whip's length. This will often result in a high SWR which cannot be tuned out and will result in an inferior performance.

By all means, follow directions enclosed with the mount you use. Most problems can be avoided by just reading the instructions. Special attention should be paid when drilling holes to make sure that the exact size drill bit required is used.

## Quickie Installations

Many motorists and lorry drivers don't want a permanent installation in their vehicle, but rather one that would permit a fast method of radio hook-up or removal. For these folks there are a few simple CB accessories.

If you have a cigarette lighter in your mobile, there is a cigarette lighter plug that you can buy, and attach to your CB radio's power cord. Then, you can remove the car lighter and simply plug into the vehicle's electrical system. Often you can get this type of plug put on your power cord when you buy your radio.

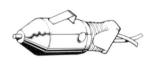

If you put it on yourself, first determine if the vehicles you'll be using are positive or negative earth. In a negative earth vehicle, the red power lead would attach to the tip of the plug, and the black lead to the side clip. A positive earth vehicle would be the other way around.

The best way to attach the wires to the plug is by soldering them on. If you don't know how to solder, there is a section later in the book that can help you out. When you are done, make sure that the red and black leads from the power cord remain insulated from each other, back where they enter the cigarette lighter plug. If you plan on switching from one vehicle to another with this set-up, make sure that they all have the same earth, or damage to the radio may result.

## Quickie Mounts

There are *magnetic mount* antennas that magnetically hold the antenna down onto any steel roof. They are great for the instant installation; you just run the coax through the window or in through the boot. A magnetic mount will withstand winds up to 120 miles per hour and will work either on the roof or the boot. Magnet mounts don't get out as well as some other kinds of antennas that are electrically connected to the chassis. But for someone on the move, they could be just the thing.

Do not cut the coax short on a magnet mount or bunch the coax up into a bundle, because the antenna will not get out nearly as well. On a magnet mount, the coax itself is part of the antenna. Also, be careful not to use a magnet mount near a luggage rack or other metal obstruction as it makes it harder to tune up. Some lorry bodies are made of aluminium, which makes it impossible to use a magnetic mount on the cab.

## Hooking Up The Coax

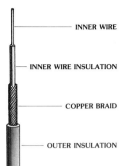

O.K. We have gotten the radio wired up. Now we have to hook up that radio to a twig. In order for the radio signal to get from the CB to an antenna, you need some kind of wire to connect them together. Usually, a special kind of two-conductor shielded wire is used, called coax (*Koh*-axe, from coaxial). This special wire is made up of an inner wire covered with a plastic sheath, and an outer wire mesh that in turn is covered by a black outer covering. Now there are several types of coax, but only two kinds are much good for CB—RG-58/U and RG-8/U coax. Unless you're running 1000 watts in your mobile, RG-58/U will work fine. RG-59/U is sometimes used for phasing harnesses.

It's good when installing a CB to get the coax up and out of the way, not only for looks, but also to keep it from getting stepped on or tripped over. Avoid flattening or crimping or wearing the insulation off.

The coax will need to be connected to the back of the radio and to the antenna. This is done with a connector (such as a PL 259) that attaches to the coax.

The coax and connector are often supplied when you buy an antenna. Sometimes, though, you may need to attach your own connector, especially if you make your own antenna. We talk about attaching connectors to coax later in the book.

## Connecting the Coax to the Antenna

Some antennas have the coax already attached. Other antennas either have a connector that plugs into the base of the antenna or two terminals which attach to the antenna mount. There are a few other ways to do this, so check the instructions that come with your antenna.

It is important that the shield of the coax be securely connected to the body of the vehicle. The shield is the larger outside braided wire on the coaxial cable. Usually, the mount of the antenna provides this connection. If you are using a mirror mount type, make sure that your mirror braces are making good connection with the body of the truck.

Sometimes if the connection is not made properly, the antenna will not work right. Although you may be able to receive somewhat, you may not get out very well. Sometimes scraping down to bare metal is necessary in order to make a good connection.

## SWR    Is your Antenna Working Right?

The length of your antenna makes a difference because the radio waves have a certain length. You must match your antenna's length to that of the radio wave. When your antenna is the wrong length, some of this power does not radiate, but is bounced back into the coax, and into the radio itself. If this *reflected power* is too high, it can cause your power transistors some problems. Besides, it's a waste of energy that could get out there and do its thing for you. So you want the least amount of reflected power and the most amount of forward power possible.

You can find out if your antenna is the right length by using an SWR meter. An SWR meter can be purchased at most radio stores or borrowed from a local breaker.

SWR means Standing Wave Ratio. Sounds rather high falootin', don't it? Well, don't let it scare you off. It's simple.

### Checking SWR

You need a short (½ metre) coaxial cable with a connector on each end. Plug in one to your rig and the other to the socket on the meter labeled TXMTR or TX. Plug the coax that goes to your antenna into the socket on your meter labeled ANT.

Turn the knob on the meter all the way down (counter-clockwise). Put the switch in the *forward* or *calibrate* position. Turn on your radio and listen for other stations. It should be working normally. Tune your channel selector to some unused channel where you won't bother anyone.

Press your mike button and without saying anything into the microphone, adjust the knob on the SWR meter until the meter reads *set*, or full scale. Then flip the switch to **reflected** or **SWR** and read the **SWR** scale of your meter.

After noting what the SWR is, stop transmitting. If the meter reads less than 1.5, your antenna is working properly. If it reads more than 2, your antenna probably needs adjustment. If it reads more than 3 or in the red, check all connections at the antenna for possible bad connections; the antenna or the centre wire of the coax might be touching the body of the vehicle. Coiling your excess coax into a small bundle can cause a high standing wave. If this happens, the coax can be coiled into long loops to avoid causing a high SWR.

When installing a trunk lip antenna, make sure you scrape the paint away on the underside of the boot lid where the little set screws make contact. A poor earth connection here is a major cause of high SWR. If you obtain a high SWR when installing mirror mounts or luggage rack type antennas, check and make sure that the mirrors or luggage rack is earthed to the body of the vehicle. A poor earth here will also give you SWR problems. Antennas shorter than about .5 metres long will probably have a high SWR or not work quite as well on some channels than others. A short antenna will not tune a broad enough range of frequencies to include all forty channels.

## *Adjusting Your Antenna*

So let's say that your SWR turns out be 2:1, and you want to bring it down. There are two ways to change an antenna's SWR: one is to lengthen it, and the other is to shorten it. You can figure out which way to go in the following manner: Take an SWR reading on channel 1, then take one on channel 40. Which channel had the higher SWR? If the SWR was higher on channel 40, you need to shorten the tip. If the SWR was higher on channel 1, you need to lengthen the tip. Move it about ¼" at a time. Check the SWR on some the middle channels and try to set the antenna so that it has the lowest SWR on the middle channels.

Most antennas have a set screw you can loosen so that the tip can slide up and down. If this does not give enough adjustment, you can clip or file ¼" at a time off the bottom of the tip and reinsert it in the coil. Make sure to reset the SWR meter every time you take a reading.

One good rule to remember when adjusting is, if you bring your hand near the loading coil or top of the antenna and the SWR goes down, the tip of the antenna needs to be lengthened; if the SWR goes up, the tip needs shortening. If you're using twins, make equal adjustments to each side at the same time.

*Don't cut too much off, now!*

If the SWR is high and you are using a transistorised rig to check it, do not hold the mike button in for more than ten seconds at a time. This is to protect the power transistors.

# Chapter Four

## Base Stations

Base stations are extremely popular because they provide breakers with *free* communications to other bases and mobile stations as well. Whether you are into base operating for the late night ratchet, emergency monitoring, or for contacts from your mobile to the home twenty, a CB set-up can be a great thing to have. Within just about every good sized community there are base operators on the air giving directions and local information, as well as keeping the community informed of who's where and what's happening.

**How about it, Test Pilot? Have you copied the Blue Leader on his way to Tower Town, c'mon?**

*Negatory, good buddy. I don't believe he's been by this twenty yet. I've been earwiggin' all morning, trying out a new base twig, c'mon.*

**That's a rog. Well, if you hear him on the channel, tell him that Screamin' Demon was looking for an eyeball, and we'll try him next time through, c'mon!**

*That's a four. We'll give you the golden numbers then. Have a safe trip. This is the old Test Pilot Base going down and gone.*

## Installing a Base Station

A base station is a little different from the mobile. It's usually bigger, and it runs on 240 volts AC, so you can plug in right off the main power. You can set it up on a desk or bookshelf—somewhere that you find easily accessible and easy to monitor. Installing it is no problem: hook up an antenna and coax, and you're in business. There are many optional base station features, but the best one to have behind you is a good solid antenna that will put it out for you.

## Using a Mobile Rig as a Base

Most base station rigs just plug right into the wall. If you plan to use a mobile rig for a base station, you will need to hook up a power supply or use a car battery and charger.

Since most mobile rigs are rather small and lightweight, we suggest that you mount the rig on a table, desk, shelf or a block of wood, to hold it down. If you plan to use a DC converter, make sure you get one with at least 3 amp. capacity at 12-14 volts DC. Also, make sure that the supply is regulated.

The DC power supply is usually a small box that plugs into an AC outlet and has two terminals where you can screw on the power leads to the radio. To hook up your radio, connect the red lead from the transceiver to the positive (+) terminal and the black lead to the negative (-) terminal. If your radio has no black lead, use a piece of insulated automotive type wire to connect from a screw on the radio's case to the negative terminal of the DC converter. Make sure there is a 2 or 3-amp fuse in one of these wires.

If you plan to use a car battery instead of a DC converter you should get something to keep it in. Battery acid can and will eat your clothes and lots of other things, including your rug. There are plastic cases sold for use with electric boat motors, available at hardware stores. Whatever you put the battery in, make sure that you have adequate ventilation. Batteries let off hydrogen gas, which, if allowed to build up can be ignited by a spark or flame, causing an explosion. Also batteries should be well up out of the reach of small children, as the acid within them can cause burns and other injuries. If you are buying a new battery, try to get a maintenance free, sealed battery. This is the safest kind to use.

Using a car battery has its advantages. If you should have an AC power failure, your base will continue to function on battery power automatically. A transistorised CB transceiver will run about a week or two under normal use on a fully charged car battery.

Hooking up the car battery and charger: It should be located within fifteen feet or so from the radio, so you don't lose too much juice running long wires.

You need:

- 12 volt car battery and case.
- ½ amp trickle charger (auto parts or dept. store)
- two insulated wires of different colours (each long enough to reach from battery to the radio)
- electrical tape
- one positive battery clamp with wing nut
- one negative battery clamp with wing nut
- an in-line fuse holder with fuse (3 amps)

Make sure you get the right wires on the right battery posts

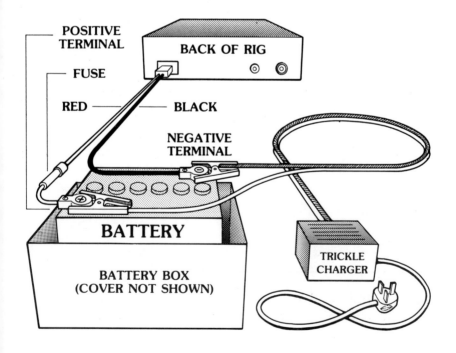

## Home Twenty Aerials

Getting yourself a good set of ears is an important aspect of getting out. You can go and buy commercially made antennas that will do a good job for you. Also you can make your own and put it up yourself—so we include some antenna building ideas later on that can save you some money and also be fun to do.

When antennas get talked about, they are called by their wavelength—like a ¼ wavelength or a ⅝ wavelength. A wavelength is the length in space of one complete cycle of any radio wave. This length will differ depending on the frequency of the radio signal. One wavelength for 27 MHz CB frequencies would be about 11 metres long. That's why the CB band is often referred to as the 11 metre band! Your antenna is made a multiple or sub-multiple of this length, because that is the right size package for catching this particular radio wave.

Driving around these days you can notice a lot of different base twigs—all kinds, shapes and sizes. The people that modulate them, talk about how well their twigs do, and you might wonder what kind of difference one could have over another as far as getting out goes. Obviously, you're bound to get out better the higher you go, so stick it on up there!

*That's right folks! With a little hot coax in the a.m., Gain perks you up and gets you out!*

Another thing you hear folks talk about is *gain*.

## Gain

The simplest antenna, a regular ¼ wave vertical, has no gain. Manufacturers call this unity gain. Other antennas that we talk about are measured by determining how many times better than this they are. Let's say that you just bought a new antenna that has a gain of 3 dBs (dB stands for decibels, a mathematical term used when comparing two signal levels). When you hook up your rig to this new antenna, you'll increase your effective power past what a ¼ wave vertical antenna could put out. A 3 dB gain is approximately a doubling of power. Here's a chart that has the gain all worked out for you.

| dB Gain | Multiply Power by |
|---|---|
| 0 | 1.0 |
| 1 | 1.2 |
| 2 | 1.6 |
| 3 | 2.0 |
| 4 | 2.5 |
| 5 | 3.0 |
| 6 | 4.0 |
| 7 | 5.0 |
| 8 | 6.3 |
| 9 | 8.0 |
| 10 | 10.0 |
| 11 | 12.6 |
| 12 | 15.8 |
| 13 | 20.0 |
| 14 | 25.1 |
| 15 | 31.6 |
| 16 | 40.0 |
| 17 | 50.2 |
| 18 | 63.2 |
| 19 | 80.0 |
| 20 | 100.4 |

If you had a new antenna with a gain of 6 dB, and your radio was putting 4 watts into it, you would have the power equivalent of running 16 watts into that old ¼ wave antenna.

One of the best twigs that you can get for a base rig has a gain of 13 dB. If you put your 4 watts output power into that, you would get an equivalent power of 80 watts in the direction the antenna is pointed. A change of one decibel in power is a just noticeable change in loudness. It takes about 6 dB to move an S-meter one unit. In other words, if you are using a regular ¼ wave and then switch to a beam that has 6 dB gain, the signal that you transmit and receive will be stronger by one notch on your and the other station's S-meter.

The gain we're talking about actually has to come from somewhere. In an *omnidirectional* antenna (works well in all directions), gain comes from building the antenna in such a way that the RF energy is spread out closer and flatter to the ground. This puts more of your power into radio waves that travel along the ground. It does this by taking power away from the higher angles. Most local communication is done on the ground wave. Some types of omnidirectional antennas are called *ground planes*.

The other kind of antenna that's *directional* is the *beam* which focuses the RF energy narrowly, by the use of several antenna elements. This gain is produced at the expense of RF energy that could have gone in other directions.

*Effective Radiated Power*

In Britain, the Home Office has limited the Effective Radiated Power, or ERP of all 27 MHz FM stations to 2 watts. The ERP of any CB station is equal to the amount of power actually reaching the antenna multiplied by the amount of gain that the antenna has. This takes into account the amount of power lost in the coax between the radio and the antenna. The attenuation chart below outlines several ways that you can lose dBs between the radio and the coax.

Four watts output power is the maximum legal power that is delivered by CB radios. If you have an in-line SWR meter, you lose ½ dB for each coax connector on either side of the meter, and 1 dB lost for the meter itself. Even a run of only 40 feet of RG-58 coax will lose you another 1 dB. With a set-up like this you would lose 3 dB, so only 2 watts of your original 4 watts would reach the antenna. If you used a high gain antenna, you could boost your signal level back up again. The Home Office decision to limit antenna length to 1.5 meters or less ensures that no gain antennas would be possible. In fact the 1.5 metre antenna has an actual loss factor of around 3 dB. This makes it quite impossible for any base station installation to even put out 2 watts of ERP, unless your antenna is mounted right on top of your base. The 1.5 metre antenna will also be less sensitive on receive than any of the longer antennas would be.

## Attenuation Chart

| Losses in dB | Source | Losses in dB | Source |
|---|---|---|---|
| -½ dB | for every coax connector in the line | -2½ dB | for every 100 feet of RG-58 between the radio and the antenna |
| -1 dB | for every SWR meter, switch, or other add-on device in the line | -1 dB | for every 100 feet of RG-8 between the radio and the antenna |

On the UHF 934 MHz FM Band, stations are allowed to run up to 25 watts ERP. Because of some of the limitations of UHF, the Home Office has allowed multi-element beam antennas for base station installations.

## Ground, Space, and Sky Waves

On 27 MHz, there are three different paths by which CB radio waves can travel: via the *ground* wave, the *space* wave and the *sky* wave.

**The Ground Wave** - CB radio signals can travel along the surface of the earth, conducted by the ground. These signals can bend around buildings and go over hills.

**The Space Wave** - when two stations are within line of sight of each other, they can communicate point to point via the space wave. If you remember that radio waves are similar to visible light, then it is clear that anything you can see, you can talk to. Because UHF frequencies are so short, 934 MHz CB radio waves can only travel via the *space* waves.

**The Sky Wave** - this is the path by which long distance (DX) communications take place. All skip signals utilize that part of the radio signal which shoots upward at the right angle to be reflected by the ionosphere back down to earth. By this path, radio waves can span great distances, from several hundred to several thousand miles. Ground and space wave signals travel equally well throughout the day or year, but sky wave conduction of radio

**The Ground Wave**

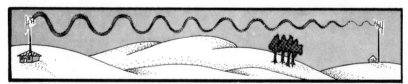
**The Space Wave**

**The Sky Wave**

waves depends on whether the sun has charged the upper atmosphere enough for it to become a reflector or not. On UHF the radio waves are so short that they never are reflected, but just pass on through into space.

*Polarisation*

Depending on the position of the antenna, radio waves travel with either vertical (elements straight up and down) or horizontal (lying flat) polarisation. Vertically polarised antennas get out via the ground wave much better than horizontal ones. That's why all CB mobiles are vertically polarised. Most base station ears are vertical or have both horizontal and vertical antennas.

A horizontally polarised antenna does not pick up a vertical antenna's signal as well as a vertical one would. But for communications between two base stations it is sometimes useful to use horizontal polarisation, because there will be less interference from mobiles. In this case, both stations need to be using horizontally polarised antennas. For DX stations, the polarity of the antenna is not important; after reflection off of the ionosphere, a radio wave can return to earth at any polarisation—it's unpredictable.

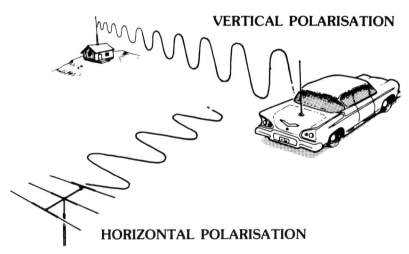

**VERTICAL POLARISATION**

**HORIZONTAL POLARISATION**

Whether you are buying an antenna or building one yourself, your first decision must be whether you want to put up a directional beam antenna or an omnidirectional vertical antenna. An omnidirectional antenna could give equal coverage for all directions locally, while a beam antenna would give you a longer range, but only in one direction at a time. You can get a TV rotator to put on that beam antenna so you can point it in whatever direction you want. Or you might possibly want both a beam and a vertical antenna for use at different times.

## The ¼ Wave Vertical Ground Plane Antennas

The most basic CB antenna is the ¼ wave length vertical ground plane. A ¼ wave ground plane has a 1 dB gain above an isotropic source, which is an imaginary antenna used for electrical gain measurements. It gets out fairly well in all directions.

This antenna consists of a driven element and three or four radials that act as a ground plane. The driven element receives the transmit energy from the rig, while the radials act as a ground. A ¼ wave vertical can be mounted very easily, and takes up very little room. It can be mounted on a lightweight metal pole or pipe such as is used for a TV antenna, or right on your roof or chimney. You can get an inexpensive "push-up" type pole, up to 50 feet tall, that a couple of blokes can put up easily.

There are other ¼ wave length antennas that are not ground planes. They usually work about the same as a ground plane. In order for an omnidirectional antenna to have any gain, it must have a driven element longer than 3 metres. Beware of ¼ wave ground planes advertising 5 dB gain!

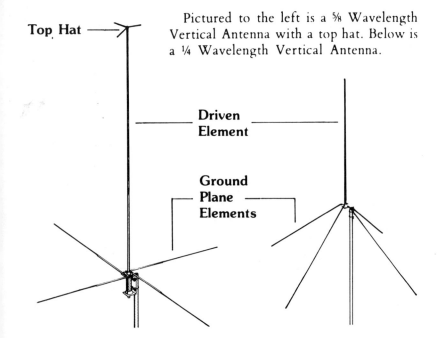

Pictured to the left is a ⅝ Wavelength Vertical Antenna with a top hat. Below is a ¼ Wavelength Vertical Antenna.

## The ⅝ Wavelength Vertical Antenna

The ⅝ wavelength vertical is similar to the ¼ wavelength vertical antenna. It too has a radiating element and three or four radials. However, the ⅝ wavelength's driven element is much longer. This is because ⅝ of a wavelength is much longer than ¼ of a wavelength. This extra length has an effect on how the radio energy is focused. This is what gain is all about—it is not an amplification, but putting to good use all the energy that is there. The main thing about a ⅝ wavelength is that it focuses its power low to the ground. A ⅝ wavelength antenna will usually have a 3 to 4.5 dB gain over a ¼ wavelength antenna. This can increase your range as much as ten kilometres or more, taking advantage of the energy that otherwise would go up and be useless, both for local communication and long distance.

There are other types of omnidirectional antennas that will work similar to the ⅝ or ¼ wavelength antenna. The best comparison in this case is by actual gain.

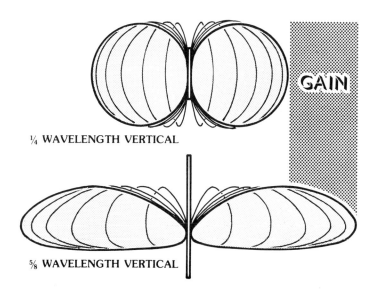

## Beams

Directional antennas, or beams, are made up of elements. An element is a length of tubing or wire positioned in such a way as to make radio waves travel in a certain direction. *Usually, the more elements an antenna has, the more gain it has, and the better it will cut out interference from unwanted directions.*

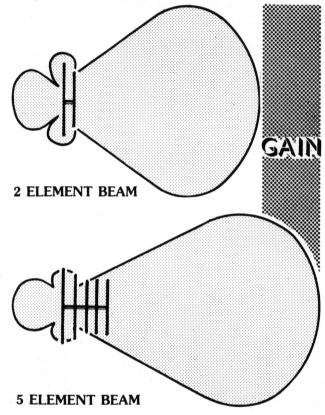

**2 ELEMENT BEAM**

**5 ELEMENT BEAM**

GAIN

There are three types of elements. The *driven element* is hooked directly to your coax cable and is the one that radiates the power from your transmitter. The *reflector element* is usually slightly longer than the driven element and is positioned 1.5 metres or more behind the driven element. It acts like a mirror behind the driven element to reflect waves towards the direction the antenna is pointing. On the other side of the driven element is the *director*, which is usually slightly shorter than the driven element. It acts like a lens to intensify the energy in the direction the antenna is pointed. Additional director elements will increase the gain.

## Vertical Beams

This is the most common type of CB beam. Vertical beams work well for both local and longer distance communication. They are usually made of lightweight aluminium tubing and need to have a rotator (a motor to turn the beam). Some of these beams are small enough to be turned by a TV rotator. Others, like the *Phased Vertical Array* are fixed in place yet change directions electronically with the flip of a switch, and can be either omnidirectional or directional.

When putting up a beam, try to keep it as far from other metal objects and antennas as possible. They could interfere with the beam's operation. Usually the vertical elements are about 5 metres (17 feet) long and are supported by a larger-diameter aluminium pipe called a *boom* and the longer the boom, the higher the gain.

Maximum Gain:

    3 element - 8 dB
    4 element - 10.5 dB
    5 element - 12 dB

## Stacked Beams

Stacking beams gives you more gain than just having one. Keep in mind that you will need a sturdy tower to hold one of these babies up, and a heavy-duty rotator to move it. The beams have to be at least 7 metres (23 feet) apart for this amount of gain, increasing the total *capture* area.

Approximate gains of stacked beams:

    3 + 3 = 11.5 dB    4 + 4 = 13.5 dB    5 + 5 = 15 dB    Mercy sakes!

## Switchable Horizontal/Vertical Beams (Criss-Cross)

These antennas combine two identical beam antennas on the same boom—one horizontal and one vertical. Each one has its own coax, with a switch to change from one to the other. Remember that on this kind of antenna the horizontal beam is not used at the same time as the vertical. An 8 element criss-cross beam has only 4 active elements

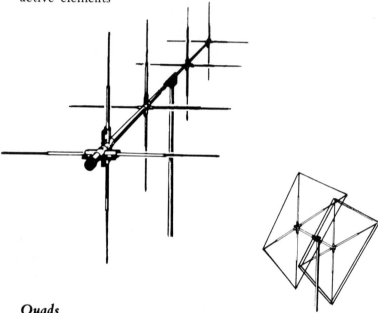

## Quads

Usually, a quad has more gain per element than a regular beam. Quads use loop or square elements, usually made of wire and supported on an X-shaped frame. They also need a tower and a rotator. Some are available with switchable polarisation and have two driven elements.

Depending on the number of elements and the spacing, the folowing gains are possible:

2 element: 8.5 dB; 3 element: 11.5 dB; 4 element: 13.5; 5 element: 15 dB.

Quads also can be stacked for more gain. Because of their large *capture* area, they are one of the best beams around. However, this area can also "capture" more wind, making it more susceptible to weather damage. If you decide to use a quad antenna, be sure that it is manufactured out of rugged, high quality, materials.

## *Coax*

When connecting your rig to an antenna, you need some kind of line to connect the two together. Now I remember once going to one breaker's 10-20 and finding that he had his antenna hooked up by using 2-lead speaker wire that ran 10 metres outside to an antenna mounted on a broom handle. Mercy sakes! He was wondering how come he wasn't getting out. The best thing to use when hooking up your rig is coax. Coax shields your line so that only your antenna radiates. Use RG-58 or RG-8 for the right impedance match. Almost all CB antennas are made to use this kind of coax.

Also its best to make your coaxial line length to your antenna as short as possible. That prevents resistance losses eating away at your power output. It starts to get critical when you run 30 metres or more of coax. You should switch to RG-8 for that long a distance, because its losses will be much less.

RG-8 coax has larger wire, which has less resistance to electrical flow than RG-58. It is also more expensive. If you transmit 4 watts through 15 metres of RG-58 cable, about 80% of the power reaches the antenna; with RG-8, 90% reaches it. If you run 30 metres of coax, RG-58 gives you about 65% of the power to the antenna, and RG-8 is about 80%.

The letter A after the number of the coax (e.g., RG-8 A/U) means that it is a new type wire that doesn't deteriorate after years of use. That is the best kind of coax to use.

There are also other numbers for coax. RG-213 A/U is the same as RG-8 A/U. Other numbers can be looked up in cable manfacturing guides.

It is a common myth that chopping off sections of your coax will help out your SWR. While this may change your SWR reading slightly, it won't necessarily help you get out any better. The best thing is to use the shortest length of coax possible.

If the SWR is high, check all the connections in the line and make sure the antenna is the right length and has been assembled properly. Try wiggling the coax at the connector while checking the SWR to see if there is a loose connection or short where it was soldered.

## Lightning

Hanging that big old antenna out there is sure attractive to lightning, and your coax will bring it right on into your house, which may produce a variety of unwanted consequences, such as a light show and Roman candle effect from your rig, and the possibility of getting zapped! Lightning usually poses no danger for mobiles. However, if you operate a base station, there are a couple of preventive measures that you can take. The best thing to do before a lightning storm is to pull the main plug and disconnect your coax connector. Make sure your coax is put away from people and things that could be zapped!

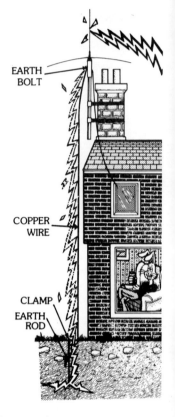

There is no substitute in lightning protection for a *properly earthed* antenna. To earth your base antenna, attach a thick copper wire to one of the mounting bolts that clamp your base twig to the supporting pole. The wire should be run directly from the antenna down to ground level without any sharp bends in it. An earth clamp attaches the ground end of the wire to an earth rod. This rod is made of metal and it should be hammered at least 1.5 metres into the ground. If you haven't got a place to drive an earth rod in that deep, you can bury 3 metres (10 feet) or more of the bare wire in a trench 6" deep. Earth rods, wire and clamps can be obtained through local electrical supply stores.

## Towers and Masts

The easiest method of mounting a base twig is to stick it up on the roof or attach it to the side of the house. One of the most popular methods of roof mounting is accomplished by using a chimney mount, like the kind used for the installation of TV aerials. There are other kinds of roof mounts available. The choice that you make will depend on what your rooftop is like. Some things to keep in mind are:

When putting it up, make sure it's in such a position that if something slips, it can't fall on a main power line.

Small antennas like ground planes or three element beams can be supported on a section of pipe or on push-up type poles with guy wires. A push-up pole is made of three or more sections of pipe, which telescope up to 15 metres or less. The sections are attached together by clamps or bolts. You attach your antenna to the top section, put the pole on the ground, and extend and clamp one section of pipe at a time, straight up. Tape your coax and any rotating cable as the pole goes up. You'll usually need a step ladder and some help from your friends.

Larger antennas, including beams, should be supported by a triangular tower. Triangular towers are available in many models. Most of them come in 3 metre sections that bolt together. Some are free standing; they don't need guy wires or support if they are connected into the ground. Check the manufacturer's specifications.

Before putting up a tower or any other large size antenna support system, it is advisable to check with the Surveyor to the local council, to see if any of the local zoning laws prohibit that kind of structure in your area. Failure to comply with local zoning laws can result in a fine and the forced removal of the antenna system.

Make sure the tower is sturdy—you may need to sink it into concrete in the ground and use guy wires to hold it up. When using guy wires on towers with beam antennas, make sure to have them attached to the tower far enough below the beam to permit rotation. Use good galvanised steel wire and an "egg" type insulator spaced every 5 feet along the guy wire until the wire reaches 30 feet down from the antenna. Past that you don't need to use insulators. This will keep them from interfering with the radiation of the antenna. Also it helps if your antenna is as clear as possible from other buildings, trees or metallic objects. Large metallic objects within 5 metres can interfere with your antenna.

Another common method of supporting a tower is to mount it against one wall of a house or building. Keep in mind that your tower may be subjected to high winds and stress.

Tower construction usually involves the help of some experienced people and a Jinn Pole. A Jinn Pole is a stout aluminium pole with a clamp assembly on one end and a pulley on the other. After sinking the first tower section in the ground, the Jinn Pole is clamped to the top of that section. A heavy rope runs through the pulley, which is now about 3 metres above the top of the bottom section. One end of the rope is tied to each new section (in about the middle). One or more people on the ground can pull on theother end of the rope, until the new section is comfortably suspended over the bottom section. From there it can be easily lowered into place and bolted to the section below. This procedure is repeated until the last section is in place.

Always use a good safety belt whenever climbing or working up on any tower.

## *Rotators*

A rotator is an electric motor that allows you to turn your beam towards any direction you want. TV type rotators should only be used with the smallest of beams, three elements or less.

For large beams, you will need a heavy-duty rotator made especially for CB or HAM use. They come with a control box which is wired to the rotator by an electric cable. Make sure to get your cable long enough to reach from the antenna to your radio, with some to spare in case you might want to move the position of your radio. The rotator is normally mounted on the tower a few feet below the top. It has a pipe called the *mast*, which goes up through a hole or sleeve on the top of the tower. The mast should be made of thick wall galvanised pipe 4.5 cm (1¾″) or bigger.

Test your rotator on the ground before you put it up on the tower. Hook it up according to manufacturer's directions. After checking to see if it rotates in a full circle, set it on north. Then turn off the power and disconnect the leads. Before tightening the final mount bolts on the antenna, make sure that your beam is pointed north—that way, your indicator and your antenna will match up. Make sure to leave enough slack in your coax between the tower and the mast to allow the antenna to rotate freely. You need to leave a little slack on your rotator cable, too, to prevent it from pulling on the connections. Both coax and rotator cables should be taped to one leg of the tower. These cables should not be brought away from the tower until you get at least 3 metres below the beam.

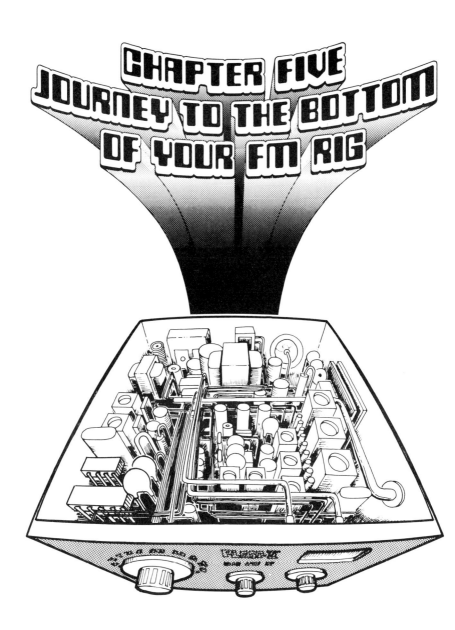

Okay, folks, it's time for a relaxed tour of your CB radio. We'll take our time, but we're not going to get bogged down in details. This "inside view" should give you a rough idea of how a radio works.

Follow me over to the antenna. Now let's grab it and slide down into the rig. Oops—watch your step around that coil; it's humming with juice! Okay, now that we're all together, everyone look down at your copy of the tour map through this section of the rig called the *receiver*.

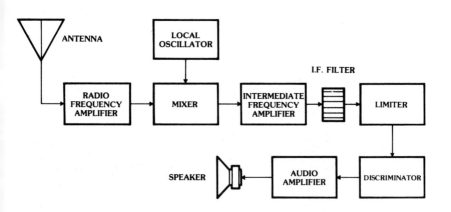

Our radio frequency slides down the antenna into a Radio Frequency Amplifier, where the signal is made a lot stronger. From less than a millionth of a volt, our signal jumps to a tenth of a volt or so.

Let's continue to follow the energy through the rig. Stay here with me; you folks are walking towards the main power supply, and there's some capacitors over there that are charged up, so be careful not to touch them. They'll knock your socks off!

With a forty channel rig, it's important to be able to select which channel you want to listen to and reject the others. There are filters that can be used that will only let a little slice of radio frequencies pass through. On a CB rig, you'd need 40 of these—one for each channel! But if we first reduce any incoming 27 MHz frequency to one standard *intermediate frequency*, then we can use the same filter for every CB channel. This frequency is 10 million cycles per second, or Hertz. That's quite a step down from 27 million. The reason for this intermediate frequency is that it helps your receiver give clearer, more selective reception.

That's the "why" of intermediate frequency. The "how" is that we run the signal through a mixer circuit, where we also shoot in another high frequency signal:

These two signals mix together and produce a third signal, just like mixing red and blue paint together will give you purple. This third frequency is the intermediate frequency and it contains all of the modulation that was within the original 27 MHz signal. The process of mixing two signals together in order to obtain a third is called *heterodyning*.

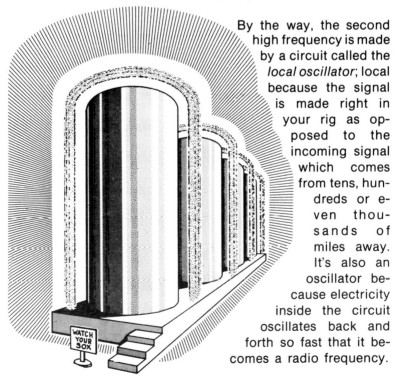

By the way, the second high frequency is made by a circuit called the *local oscillator*; local because the signal is made right in your rig as opposed to the incoming signal which comes from tens, hundreds or even thousands of miles away. It's also an oscillator because electricity inside the circuit oscillates back and forth so fast that it becomes a radio frequency.

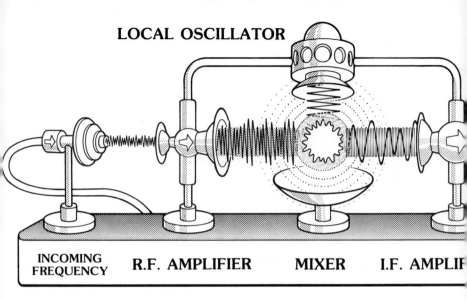

INCOMING FREQUENCY | R.F. AMPLIFIER | MIXER | I.F. AMPLIF

So, now we have a much slower signal coming out of the mixer, at usually 10 million Hertz. Once again we make the signal stronger by running this frequency through an I.F. (intermediate frequency) amplifier and filter, helping to select just the frequency we want.

We're about halfway through our receiver now. If any of you want to rest, you can sit down on those resistors over there. Warm, ain't they? That's because some juice goes through them and resistors just use juice up as heat. So, get comfortable while I tell you about the next mindboggling circuit!

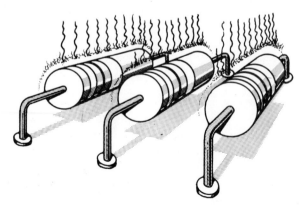

FM signals are quite different from AM ones. Although an FM radio wave is always kept at an even strength, its frequency

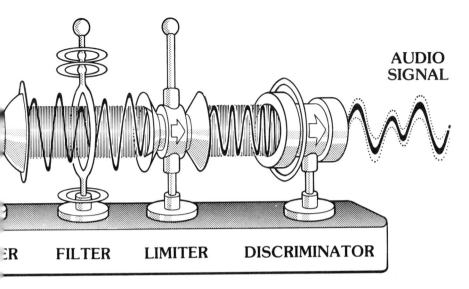

continually shifts in step with the audio frequencies produced by your voice.

That's why it's called Frequency Modulation. With AM (Amplitude Modulation), however, the original voice frequencies are imprinted onto the radio wave by varying its strength or amplitude. Since most static and noise impulses are also signals that vary in amplitude, AM receivers are prone to noise interference.

A good FM receiver will severely reduce or eliminate all amplitude varying signals by clipping off the peaks at the top and bottom of the I.F. signal. And that's what this next little beauty does: it's called the *FM limiter*.

The next circuit on the tour is called the *discriminator* and its job is to take the audio signal off of the I.F. frequency after it has gone through the limiter. The audio is contained in the I.F. signal just like it was in the original radio signal that came in the antenna behind us. We reduced the incoming signal to ar intermediate frequency, but that didn't affect the voice frequencies at all. The discriminator has the ability to turn the frequency shifting of the FM signal back into audio. It's at this point that we discard the radio frequency energy. The radio signal brought the voice through the ozone, but now that we got it, we have no further use for it.

That's why the radio frequency energy is called a *carrier*—because the voice is the information, and once it is delivered,

the carrier has served its purpose. It's like when you bring home a pizza from the take-out place: it's the goodies that you're interested in, not the container.

Coming out at the far side of the discriminator is a *voice signal*, just like when it left the mouth of the person transmitting to you. We then run this audio signal through an audio amplifier or two so it's comfortably loud, and then it goes right into a speaker where the signal is turned from electrical waves back into sound waves that we can hear. Now before any of you go slipping out the speaker and onto the floor, let's turn and go back into the radio, and find out how this contraption works.

Everybody rested up from going through the receiver? We're actually over halfway done, because some of the circuits we've walked through do double duty in both the transmit and receive parts of the trip.

See that big plastic container over there? That's the *relay*. The relay is a kind of switch which connects either the transmit or the receive circuits together. It's controlled by the push button on the microphone. That's how the parts common to both transmit and receive are switched back and forth. After all, transmitting is just receiving in reverse.

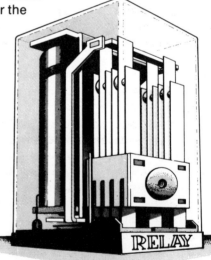

Okay, everybody. Let's put an eyeball on our tour map so we all know where we're going. Let's stay together and not get lost through all these twists and turns.

It doesn't matter how much we amplify a voice signal, it just won't radiate off your antenna, it's too low a frequency. That's why we need a radio carrier, and we'll see how it's produced in a minute.

I'll have to ask you kids over there not to spill your soft drinks on that circuit board—you'll make everything sticky and the guy who owns this rig we're walking through won't know what's

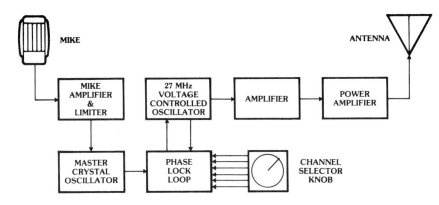

happening the next time he keys down!

If you'll look where I'm pointing, that's where the Master Crystal Oscillator is located. That square tin can over there contains a sliver of quartz crystal which physically vibrates, generating an electrical frequency.

A crystal is just what it says. It is a piece of quartz crystal (a rock) in a can. It operates on the same principle as a tuning fork. When you hit a tuning fork, it will vibrate at a particular frequency. The tone or frequency depends on how the tuning fork is constructed. A crystal operates in a similar way. When hit with electricity, the crystal will vibrate at a frequency, which is determined by the thickness of the crystal.

If you had to buy a crystal for every single frequency needed in a 40 channel CB, it would cost you more than the price of an entire radio. Some of the older American CBs used a few crystals and ran the frequencies they produced through some mixing circuits to get 23 channels. But with the electronic industry's ever growing demand for quartz crystals, even this arrangement became undesirable, and radio engineers came up with a new idea.

There is another kind of oscillator which *can* be tuned over a wide frequency range. It generates frequencies by means of a *tuned circuit* intead of a crystal. The parts inside the tuned circuit cycle electricity back and forth, vibrating at the 27 MHz carrier frequency. This oscillator's frequency can be altered by changing the voltage at one part of the tuned circuit. That's why this kind of oscillator is called a *voltage controlled oscillator* (VCO).

By itself, the VCO isn't stable enough to be the generator of a stable radio signal. It drifts away from its set frequency too easily—something undesirable in a radio transmitter. Fortunately, a VCO can be combined with some other circuits which help to keep it anchored onto the channels. This combination of circuits is called the *Phase Lock Loop* (PLL).

With the aid of the crystal oscillator and the VCO, the Phase Lock Loop can easily generate all 40 CB channels. Instead of more crystals, the PLL uses several inexpensive integrated circuits—by-products of the computer revolution. Each integrated circuit (or IC) is actually several transistor circuits contained in one small plastic blob.

Hey, how about one of you kids going over to the channel selector and flipping us onto channel one!

The channel selector knob programs one of these IC chips to divide the crystal frequency down to a fraction of its original frequency. The Phase Lock Loop compares this signal to a sample of the VCO frequency. If the PLL senses no difference between them, then the VCO is locked onto channel one's frequency. If the VCO starts to drift off frequency, the PLL's comparator will sense the change and send a correcting voltage back into the VCO. Once back on track, the difference between the two signals disappears, and the PLL shuts off the correcting voltage. The more that the VCO has drifted off of the programmed frequency, the more voltage required to bring it back into line.

So let's flip over to channel five and see what happens. Every time that we switch channels, we are making a small, calculated change in how the master crystal frequency is divided. This unbalances the compared frequencies just enough to make the

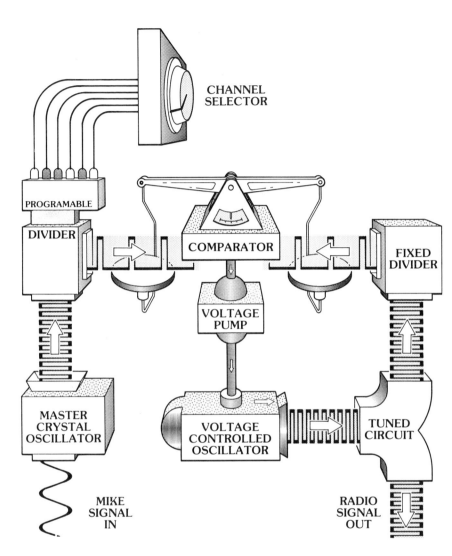

PLL readjust the VCO on up to the new channel's frequency. It will remain locked to the new channel's frequency until we spin the knob again. It is the channel selector that sends a coded message to the PLL causing this whole chain of events to occur. So as we've seen, this is some kind of fancy circuit. By using a crystal as a reference frequency, the PLL can make a normally finicky VCO have the frequency stability of a precision crystal!

Well folks, before we follow this VCO's 27 MHz frequency on up towards the antenna, let's take a look at how the audio signal from the microphone gets added onto that carrier frequency.

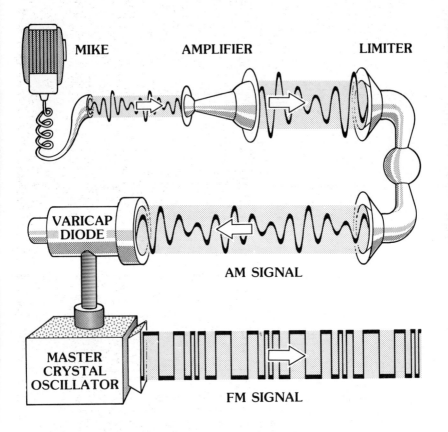

Although a crystal is normally a highly stable component, it is possible to vary its frequency slightly by hooking up a capacitor to it. There is a special electronic part called a varicap diode, and it will act like a variable capacitor when a changing voltage is applied to it. In order to get FM, the mike's audio is amplified and then hooked up to the varicap diode. The amplified audio signal causes the varicap to swing the crystal's frequency in step with your voice. And because the crystal's frequency is directly linked by the PLL to the carrier frequency, the modulation is transferred on. From there, the modulated carrier is amplified on up to a power level of several watts and is connected to the antenna plug.

Well, here we are again at the antenna. I hope none of this has left you out in the street. We've kind of gone all the way through this maze and come all the way back round to the beginning. That power amplifier was the last circuit.

Now everybody get ready. We're getting out on the antenna now and we're about to leap out there into the sky. We'll radiate on out there, traveling at 186,000 miles per second for thousands of miles out into the Universe, ending up heaven knows where.

# CHAPTER SIX

## FM, AM AND SSB

FM, AM and SSB are three ways that radio waves can be altered to carry your voice. There are several differences between these three modes of communication that you should know about. There has been a lot of controversy surrounding the British government's decision to legalize FM CBs only. Each mode has its own good and bad points. We want you to understand what they are and how they affect your ability to communicate.

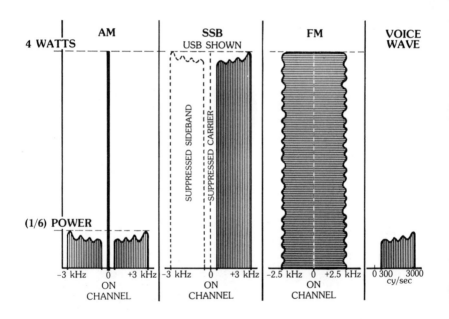

## *Frequency Modulation*

Let's start out by examining an unmodulated radio wave or carrier. Both an AM and a FM carrier are exactly the same. It's only after the addition of your voice that the FM and AM radio waves take on their own special characteristics.

The strength of an FM radio's signal or its power output always remains the same. What changes when we modulate the carrier is the frequency of the radio wave. The higher the tone of your voice, the more times a second that the FM carrier will swing back and forth from the channel frequency. If you hummed a 1000 Hertz tone into your mike, the radio wave will change frequencies 1000 times in one second.

## *Deviation*

The total amount of frequency change or swing from the carrier frequency is called the *deviation*. Legally, the maximum change allowed from the 27 MegaHertz channel frequency is ± 2.5 KiloHertz, and all FM transceivers are preset at the factory to prevent you from going over that. So if you were on channel one, or 27.60125 MHz, the carrier could swing as high as 27.60375 MHz or as low as 27.59875 MHz. If an FM transmission did deviate too much, it would exceed the FM CB receiver's bandwidth and would sound distorted. It could even overlap onto the next channel, causing interference.

With FM, the louder the voice going into the mike, the greater the deviation. And the greater the deviation, the louder that the audio will be when it comes out of the receiver's speaker.

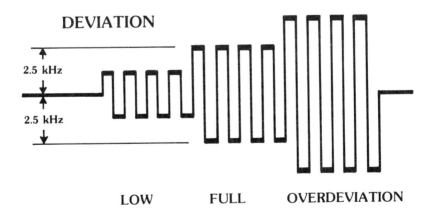

Other FM radio services are allowed to use wider deviations. Two-way radios in the VHF bands are used by police, ambulances, fire departments and other emergency services because of FM's high fidelity and solid sound. All these radios use a ± 5 kHz deviation. FM broadcast stations use a ± 75 kHz deviation. Because of music's wider frequency range (up to 20,000 Hz), a wider deviation is necessary in order to reproduce the full spectrum of audio dynamics. But this also takes up bigger channel spaces in the radio bands. Most two-way radios are designed only to hear and transmit sound in the lower spectrum of human hearing. They won't pass sound frequencies higher than 3000 Hz. This applies to most AM and SSB radios as well.

Using a narrow deviation of ±2.5 kHz makes it possible for an FM CB signal to occupy slightly less room per channel than an AM signal would. This narrow deviation, however sacrifices some of the signal's voice power.

## The Capture Effect

One unique aspect of FM is called the "capture effect." An FM receiver will be captured by the strongest signal on channel. Any other weaker breakers on channel at the same time will be overridden! In some ways this limits the range of an FM signal in cities or other crowded areas where there is a lot of CB activity. If you are trying to copy a weak mobile when some strong station keys up on channel, it's all over. You'd have to wait until the big station stops transmitting in order to hear the weaker station.

But in other ways, the capture effect can help extend the communications possibilities. With FM, there is no heterodyne or squeal as there is when two AM stations key down together. Stations on one side of town can have a ratchet with their neighbours, while a completely different set of conversations can be happening on the same channel on the other side of town. That's because stations only hear those breakers who are close to them, unless the channel becomes quiet in their area first. A signal only has to be a little stronger than the background noise in order to capture the FM receiver. When captured, an FM receiver is quieted—background noise on the channel is no longer heard. This makes for a clean copy.

*Kicking Out The Noise*

Another aspect of FM is the ability of a good receiver to reject static and other noise impulses. Most noise is similar to AM in that it varies in strength rather than frequency. A good FM receiver will not respond to signals that vary in strength. Consequently, neither noise or AM modulation are copied on an FM receiver. Also FM receivers offer a more effective squelch control than an AM radio does. **Generally, if a station isn't strong enough to break the squelch, then it wouldn't be good enough for you to copy anyway. So you should keep your FM receiver squelched, eliminating background noise.** Both of these noise eliminating factors put FM a notch or two above the normal racket that is an unavoidable part of AM/SSB listening.

Once a mobile FM station gets close enough to you to fully quiet the background noise, any further addition in signal strength will make little difference in the receiver's loudness. Once you have set the volume control to a comfortable level, it is usually unnecessary to readjust it. Also FM signals will not overload an FM receiver—even if the other breaker is parked right next to you. An AM receiver's front end would become overloaded in this kind of situation, causing distortion of the received signal.

One of the main complaints made by AMers about FM CB is that they can't talk skip on FM. This is true only because most foreign DX stations do not have FM capabilities on their CB rigs. FM signals will skip however, and the quality of skip on FM is about the same as AM. Generally, FM will get you out as much or more than AM, but SSB can take you even farther.

## *Amplitude Modulation*

When a radio wave is amplitude modulated (AM), the carrier frequency never varies. It's the strength or amplitude of the carrier that is continually changing. The louder the sound reaching the mike, the greater the variation of the carrier's strength and the louder the audio will be at the receiver. The pitch or frequency of your voice controls the rate that the AM carrier varies in amplitude. If your voice is vibrating at a frequency of 1000 Hz, then the strength or amplitude of the carrier will change 1000 times a second.

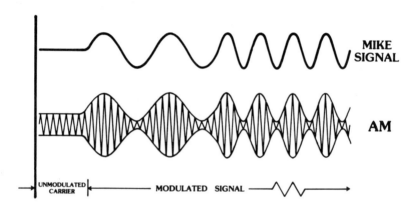

The percentage of modulation is the amount of modulation riding on the carrier. One-hundred percent modulation is the heaviest change you can put the carrier through without overloading it.

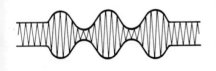

**50% MODULATION**

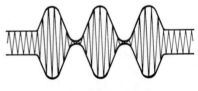

**100% MODULATION**

## Overmodulation

An overmodulated AM carrier is still understandable, but it will cause bleedover onto nearby channels and transmit spurious signals or *harmonics* which can cause interference to other radio services throughout the higher radio bands. AM CB radios are manufactured with an automatic modulation limiter which prevents most units from overmodulating. But in some situations (like when a power mike is added to the radio), overmodulation can be a problem.

## AM Noise Elimination

An AM receiver is constructed to respond to signals that vary in amplitude. Since most static and noise impulses are amplitude changing signals, they are received along with the CB signals. The development of effective noise blanker and automatic noise limiter circuits have done much to improve the quality of AM reception. But they require manufacturers to employ more complex electronics inside their units. Many of the cheaper AM CB radios do not have effective noise eliminator circuits.

## AM Receiving

Unlike FM, an AM receiver is able to copy weak signals that are right down in the middle of the background noise. An FM receiver would be "captured" by the noise and not the signal in this kind of situation, so no copy would be possible. When more than one AM station keys up on the channel simultaneously, it is still possible to copy both stations on AM with the accompaniment of the squeals or heterodyne caused by the two signals beating together.

An AM receiver is also capable of picking up FM transmissions occurring on one of the channels, and an understandable if somewhat thin sounding copy is possible.

## Single Side Band

When the mike signal and carrier are mixed together in the modulator, they form a new signal which is the composition of the two signals. The audio is both overlaid and underneath the carrier frequency. These two layers of signal are called the upper and lower sidebands.

At 100% modulation of an AM carrier, two-thirds of the total power is contained in the carrier, and one-third is divided between the sidebands. The sidebands are like mirror images and either one has all the audio signal necessary for reception. By filtering out the carrier and one of the sidebands on transmit, we can have a stronger, more compact signal. SSB transmission puts all of the power into the message. That's why an SSB signal has a gain of 8 dB over an equivalent AM signal. Switching from AM to SSB can more than double your range.

Most HF two-way voice transmissions around the world use SSB now instead of AM or FM. Because of the lower bandwidth, more SSB stations can fit into the high frequencies that are shared by worldwide radio services. In the CB Band, SSB doubles the amount of channels for use, making more room for everybody.

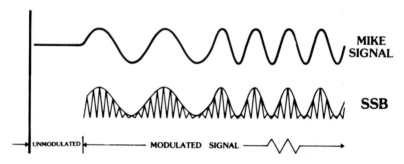

In order for SSB to be received as an understandable audio signal, the missing carrier must be added back into the SSB signal at the receiving end. That is why SSB signals sound so muffled and jumbled when heard on an AM receiver. SSB receivers use an additional circuit, which adds the carrier back in. This circuit is called a Beat Frequency Oscillator (BFO).

For a faithful reproduction of the transmitted audio, it is necessary to use a fine adjustment called the clarifier. This allows you to tune in to the exact frequency of the received station. The clarifier does not control your transmitting frequency or otherwise affect your outgoing signal. Because of this extra control, SSB radios are harder to operate than AM or FM CBs. For an understandable reception, you must select the correct sideband of the station you are listening to. More natural sounding voices are possible on AM or FM, in contrast to the duck-squawking heard when using SSB. On an SSB receiver it is possible to copy AM signals by "zero beating" the carrier. SSB receivers will not copy FM signals.

Some SSB radios have VOX or voice operated transmit. Instead of keying the mike for every transmission, just you talking into the mike will key the transmitter up. During pauses between phrases or sentences the transmitter will unkey, giving the other station an opportunity to break into the conversation to make comments. A VOX operated two-way QSO is much more like a telephone conversation than the key down FM or AM contact is.

SSB has a greater range per watt over FM or AM and is the favourite mode among skip talkers. SSB has the ability to cut through static and noise, when an AM or FM signal wouldn't make the trip. And SSB's efficient use of the frequency space makes it more appealing for use in cities and other congested areas. But SSB requires a much more complex and expensive circuitry to compact and filter the radio wave. SSB is not too compatible with FM or AM. When FM or AM and SSB signals co-inhabit the same channels, they can clobber each other, causing mutual interference.

# CHAPTER SEVEN

## DX Guide For CBers

So go ahead England! You got the Momma-Mia in the pizzaria, c'mon!

For many years, Short Wave Listening (SWL) has been a popular hobby of hundreds of thousands of people throughout the world. Many folks depend on Short Wave Broadcast Stations for news, weather and other important information. Others tune in just to copy the fascinating spectrum of activities occurring throughout the 3 to 30 MHz HF bands. It is possible to monitor ships at sea, airplanes, radio amateurs and military and governmental frequencies. But we think that SSB and AM skip contacts from around the globe can be as interesting as any of the above. If you want to get a copy on some of this skip action, then follow us out to DX-Land and we'll let you in on when and where to listen.

### Sunspots

Violent whirlpools of ultraviolet radiation located on the surface of the sun are responsible for the frequent reception of CB stations from around the world. These solar storms are called sunspots, and the intense radiation released by them electrically charges the upper layers of the earth's atmosphere, called the *ionosphere*.

### The Sunspot Cycle

Although sunspot activity has been monitored and recorded for over 200 years, it has only been associated with radio since 1901, when Marconi first bridged the Atlantic with radio waves. Scientists then theorised that sunspots created an electrically charged region in the earth's atmosphere which could reflect radio signals across the ocean.

Now we know that the number of sunspots goes up and down in an eleven-year cycle. The last peak was in 1980 and the next one should occur in 1991. We are still learning about sunspots and there may even be other, longer term cycles of solar activity that we don't know about yet. But one thing that we do know: the more

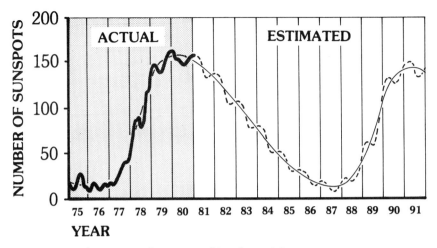

sunspots there are, the more skip there is!

Although we have passed the peak of the current sunspot cycle, there will still be good conditions for the next couple of years or so. By the mid-1980s the cycle will reach its low point and much of the time the CB band will be dead except for local communications. Then a new cycle will begin and the rising number of sunspots will cause the skip activity to pick up again. By 1991, at the very peak of the next cycle, skip conditions will exist throughout the day. During the right times just about every part of the world can be heard!

Radio waves mainly travel in straight lines like beams of light. Although some parts of a radio wave will travel along the surface of the earth, other parts will shoot off at various angles into the sky. When a radio wave reaches the ionosphere it can bounce off of it much like a mirror reflects a beam of light. The signal is reflected back down to earth, traveling from 350 miles to several thousand miles. Sharp angles may not be reflected; consequently, there is a

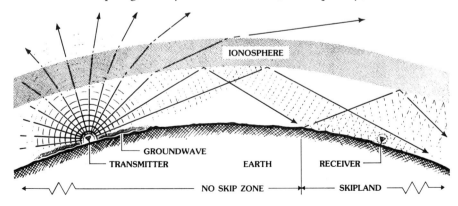

zone where communication is not normally possible. On other occasions it is possible to talk into this "no skip zone."

The ionosphere is actually made up of several layers that are located anywhere from 30 to 300 miles above the surface of the earth. The height of the lowest layer at any given moment depends on the season, the time of the day and the intensity of the sun. When the layers are not very strongly charged, the radio waves may pass through and not be reflected back to earth. If it is strongly charged, the signal may bounce once, hit land or water, and bounce back skyward to be reflected again by the ionosphere. This kind of skip is called multiple hop, and by this method, radio waves can span the globe when conditions are right.

## *The Seasonal Skip Cycle*

In addition to the eleven-year sunspot cycle, the year's seasonal changes affect the ionosphere for the same reasons that the earth's revolution around the sun affects the warmth and coolness of the earth.

***Summer*** – during the summer months, the earth and its atmosphere reach much higher temperatures than at other times of the year. The heat causes the ionosphere to expand, lowering its overall charge intensity.

At the height of the sunspot cycle summer provides enough ionisation to allow reflection of radio waves, although the skip distances are often not as long as at other times of the year. Since the summer nights are shorter, there is insufficient time for the ionosphere to discharge between days. So there are less intense variations in the skip throughout a 24 hour period than there are at other times of the year.

Summer during the medium to low years of the sunspot cycle is a different story altogether! The ionosphere will often be insufficiently charged for much regular skip to happen.

***Winter*** – during the winter months, the cooler atmospheric temperatures cause the ionosphere to contract, which gives it a more intense charge. This causes excellent long distance skip during the day. January and February are the best months in the Northern Hemisphere for DXing. The skip is often very long, with band openings to particular far away areas occurring reliably every day.

The longer winter nights give the ionosphere plenty of time to discharge between the periods of daylight. Consequently, the nighttime skip is much more unreliable than in summer, with frequent periods of low skip activity that can last for days or weeks on end.

***Spring and Fall*** – these are transitional periods, with the early spring and late fall exhibiting a tendency toward the winter conditions, and the late spring and early fall somewhat closer to the summer conditions. During this period, October and November are the best months for DX.

## *The Daily Skip Cycle*

As each day rolls on, different parts of the world will become accessible for DX contacts. Often the skip is at its best when the sun is about halfway between you and the area you want to listen for. This is because the sun has sufficiently charged the ionosphere between the two stations. You will usually hear skip from the east in the morning, and from the west in the afternoon. North/South skip can happen at any time of the day. Around the times of sunrise and sunset the skip is often the longest. The following chart should give you a general idea of when to listen to your favourite foreign stations. But remember that skip conditions are not totally predictable by any means. Even during the low point of the sunspot cycle there can be isolated peaks of skip activity that open up without warning!

***Predawn*** – skip openings occur sporadically until shortly before sunrise.

***Sunrise*** – at this time the sun has charged the ionosphere for thousands of miles to the east, making it possible to have very long skip in that direction until mid-morning or so.

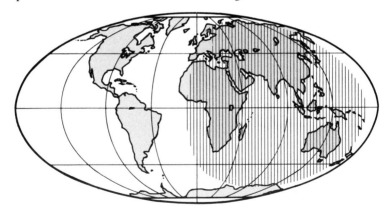

*Mid-Morning*

***Mid-Morning*** – skip to the east becomes somewhat shorter and North/South contacts are possible.

***Midday*** – with the sun overhead, skip contacts in all directions are likely. This is a good time for skip of 800 miles or less into Europe. Usually the very long skip to the east has tapered off by now.

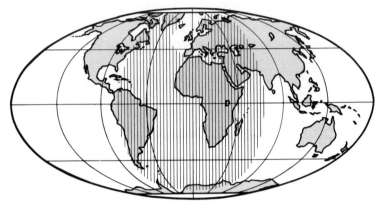

**Midday**

***Late Afternoon*** – longer skip into the west begins now. Skip from the east drops off, but North/South path is still open.

***Sunset*** – at sunset skip to the west can be open for very long distances. North/South contacts are still likely.

***Evening*** – during most of the year, skip conditions taper off within an hour or two after sunset.

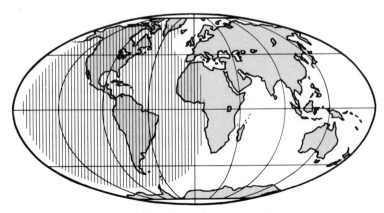

**Late Afternoon to Early Evening**

## Solar Storms and Radio Blackouts

Solar flares are magnetic disturbances on the surface of the sun which sometimes accompany heavy sunspot activity. Solar flares emit streams of sub-atomic particles that can totally disrupt most High Frequency skip for periods ranging from a few hours to several days long. These particles reduce the ionosphere's ability to reflect radio waves. As these particles bombard the earth they are attracted to the north and south magnetic poles. Their passage through the high polar atmosphere can actually ignite the gases there, creating aurora displays.

As the sun rotates, so does the storm and eventually it disappears to the other side which faces away from the earth, allowing normal conditions to return. But 27 days after the radio blackout first occurred the storm can reappear again from the other side of the sun and once more exert its devastating effect on CB as well as other HF skip communications.

## Fading

On some occasions, a transmitted radio wave may take more than one path to any particular location, causing there to be a slight time delay between the two signals. When they recombine in your receiver the two signals may cause the reception to get stronger, weaker or fluctuate up and down in an erratic manner. These changes in received signal level are called fading.

## Long Path

On many occasions, the shortest physical distance between two stations may not be the actual path that a radio wave takes when skipping. *Long Path Skip* occurs because there is a strong enough charge on a section of ionosphere that stretches the long way around the earth. *Long Path* openings to Australia, New Zealand and the South Pacific happen frequently and usually occur at different times of the day than the regular openings do. DX stations with beam antennas must keep this in mind, because it can be quite critical if a beam is pointed in the opposite direction during a long path opening!

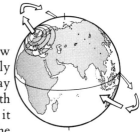

## North/South Skip

Sections of ionosphere over the equator and the surrounding regions tend to get a higher charge than do other areas. This provides a path for DX contacts between the earth's Northern and Southern Hemispheres, even when skip is impossible in other directions. During the low sunspot years most skip will be possible in the southern hemisphere rather than east/west. North/South skip is also more common during the summer months. Listeners in Europe will tend to hear a lot of South American and African stations within this period. (Habla Espanol, amigo?)

## Sporadic E and Other Exotic Skip

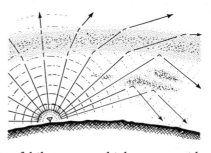

*Sporadic E skip* – is caused by irregular, low-layer clouds of ionised air. The phenomenon occurs most often in the late Spring and Summer months. There can be a small fast moving cloud reflecting radio waves from one isolated location to another or there can be large clouds covering hundreds of kilometres which can provide multiple hop skip. Sporadic E occurs frequently in the mid-morning and late afternoon, although it can occur at any time of the day. During the low part of the sunspot cycle, there are periods of time when this is the only kind of skip happening on 27 MHz.

*Aurora Skip* – happens when signals can be reflected off the aurora borealis. Aurora is usually accompanied by visual displays but there can also be radio aurora without a visual display. Sometimes auroras cause polar blackouts and radio communications over the poles fade out. When conditions are favourable, you can point your beam antenna north and hear other stations by banking off of the aurora borealis.

*Back scatter* – can happen anytime of the the year. All the stations just point their beams south and their signals bounce off ionised clouds near the equator. With an almost direct ricochet you can talk to nearby stations which would normally not be open to you. Your signal may also split off (side scatter) at a greater angle providing communication as far away as the States. This usually results in weaker signal levels, so usually only stations with beams can work scatter.

*Heat Inversion* – when this occurs, a radio wave can travel around the curvature of the earth, spanning greater than normal distances on the ground wave. Heat inversion usually happens in the summer, when hot and cool layers of air collide, creating an electrical charge between the layers. This static charge can cause long ground waves. It is usually noticed at night when the rest of the skip disappears.

## Burners

With the number of CB radios in use today, there's lots of people transmitting on all 40 American channels at the same time. So when the skip rolls in, usually there's a lot of noise that comes from so many stations being on the air at once. The squeals, whistles and bumblebees you hear on AM when two or more stations are transmitting on the same channel at once are called heterodyne. The only way to be heard above the heterodyne is to get louder than the rest of the stations. Most operators that you hear will be using very large antennas and illegal power amplifiers (linears, shoes, burners, etc.). Linears are hooked in the coaxial line between the radio and the antenna. Most of the mobile burners are all solid state and operate off of 13.8 volts DC. These units produce from 30 to 300 watts depending on the model. Base amplifiers run off of 240 volt AC mains and can be valve or transistor units. The higher power base units (up to a kilowatt or so) are usually valve units. The use of power amplifiers increases the strength of a DXer's signal. It also increases the amount of interference given to other radio services and electronic equipment.

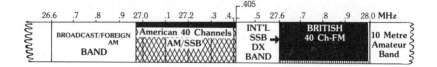

## The Eleven Metre Band

In most of the world the skip talkers use extra frequencies besides the regular 40 American CB channels. There is so much noise from the millions of CB users worldwide, that only the most powerful stations can work skip on the regular AM channels. Many of the skip talkers operate above and below these channels. In the U.S. and elsewhere, these operators are called *Outbanders*, and many of them are using sliders, VFOs or even converted Ham Radios. With this kind of equipment, CB DX operators can move freely throughout the 11 Metre Band.

The Home Office legalised CB on frequencies up above the US 40 Channels, to prevent American and other foreign skip from taking over the British FM channels. But the new frequencies that were chosen have been used for years by thousands of SSB DX stations around the world. It's only a matter of time before some of those stations start talking on FM, especially when they start hearing all the English breakers skipping in!

## QSL Cards

Many DX stations have cards that they send to one another to confirm their radio contacts. These cards are called *wallpaper* or QSL cards. The information usually contained on these cards include: the operator's call numbers or handle, name and address, time and frequency of contact, a signal report that describes the copy, and the type of radio and antenna in use. Many QSL cards are personalised with artwork and sayings that pertain to the particular station. If you hear someone coming in good from DX Land, jot down their address and drop them a line. Most folks are glad to know they are getting out and they'll send you back a card. Short Wave Listeners have been doing this for years, getting QSL cards from foreign broadcast stations.

## International 40 Channel Band Plan

| Channel | Use |
|---|---|
| 1, 2, 3 | FM |
| 4, 5, 6, 7 | open |
| 8 | guard |
| 9 | emergency |
| 10 | guard |
| 11 | breaking |
| 12, 13 | open |
| 14 | breaking |
| 15 | open |
| 16 | SSB only |
| 17 | calling |
| 18 | open |
| 19 | travelers |
| 20, 21 | open |
| 22, 23, 24 | open |
| 25, 26, 27 | calling |
| 28, 29 | AM only |
| 30 | SSB calling |
| 31, 32 | open |
| 33, 34 | FM |
| 35 | FM breaking |
| 36 | open |
| 37 | FM calling |
| 38 | SSB calling only |
| 39 | SSB only |
| 40 | open |

There are stations all over the world operating AM, FM and SSB on the 40 American Channel frequencies. It is important to use the channels designated for your particular mode, in order to keep from clobbering stations using the other modes. Also certain channels are used for specific purposes internationally including 9, 11, 16 and 19. *Channel 11* is used as the AM breaking channel internationally. It is the breaking channel in the Republic of Ireland. *Channel 14* is the breaking channel in London as well as some other areas of the UK. *Channel 16* is used in most parts of the world as the SSB breaking channel. Usually you can find many stations on this breaking channel trying to talk with other CBers outside their native countries.

*Guard Channels* should be kept free as much as possible to prevent bleedover interference onto *Channel 9*, the international **Emergency Channel**. *Calling channels* are used by CB clubs and other groups as a means of staying in touch with fellow members. Channel 38 is an international SSB calling channel.

# SSB LINGO

*Calling CQ, CQ, calling CQ DX and beaming North America, this is Zulu Lima four-oh-six, in South Auckland, New Zealand calling CQ DX and standing by.*

*Zulu Lima four-oh-six, this is Echo Bravo three-seven-niner in southern Ireland. Do you copy? Over!*

**Roger Roger. Echo Bravo three-seven-niner, we have a five by nine copy on you this afternoon in New Zealand! The personal on this end is Mike, and my QTH is 15 kilometres southeast of Auckland, over.**

*Roger Mike! Excellent copy on you here. You are five by nine plus twenty. The handle here is Richard and we are running a Yaesu FT-101 Double E here into a five element Long John Antenna. So back to you Mike and we'll see how conditions are holding up. Zulu Lima four-oh-six, this is Echo Bravo three-seven-niner, standing by.*

SSB DX operators talk a whole other language than the AM CBers. Instead of using handles and CB lingo, SSB operators talk in a way similar to radio amateurs. They use SSB club call signs and their personal name instead of a handle. SSB operators also use the internationally accepted Q-signal code, instead of the 10 code. The Q-code was originally developed by radio telegraphers as a kind of radio shorthand to save time when sending lengthy messages via Morse code. This practice has carried over into use by radio amateurs and SSB CB operators. A list of the more common Q-signals appears below:

## International Q-Signals Used by HAMS, SSBers, and Skip Talkers

| | | | |
|---|---|---|---|
| QRA | Name or handle | QRX | Call back later, stand by |
| QRE | Estimated time of arrival | QRZ | Who is calling me? |
| QRH | Frequency varies (FMing) | QSA | Readability |
| QRG | Exact frequency | QSB | Fading signal |
| QRL | Busy | QSL | Acknowledge receipt |
| QRM | Interference from other stations | QSM | Repeat the message |
| QRN | Natural interference-static | QSP | I will relay to _____ |
| QRO | High power | QSO | Communications with, contact |
| QRP | Low power | QSX | Listening on channel _____ |
| QRQ | Transmit faster | QSY | Change frequency |
| QRS | Transmit more slowly | QSZ | Send each word or sentence more than once |
| QRT | Stop transmitting | | |
| QRU | I have nothing for you | QTR | Correct time is _____ |
| QRV | I am ready | | |

CQ is another old time radio expression that lets other radio operators know that you are looking for a contact. CQ DX means that you are only looking for skip and not local contacts. Single sideband operators usually talk about the frequency rather than the channel that they may be on and they tend to avoid using CB lingo or jargon in their conversations.

## Readability and Signal Strength Reports

Besides swapping QSL cards, SSB DX operators usually wish to exchange signal reports. This is done by using the R-S Report Code that has been in use for many years by HAMs and commercial radio operators.

## Phonetic Alphabet

Another aid to the exchange of information via the radio, is the phonetic alphabet. Sometimes a station will want to get his name or address through to another operator when conditions change for the worse. Perhaps he wants a QSL card from some breaker on the other side of the world! By spelling the words phonetically, he can increase his chances of getting through the QRM and QRN. For example, Chicago, USA would be sent as "CHARLIE - HOTEL - INDIA - CHARLIE - ALPHA - GOLF - OSCAR, UNIFORM - SIERRA - ALPHA."

### R-S Reports

*Example: Your signal is coming in 5 by 9 here.*

*Readability*

1 – Unreadable
2 – Barely readable
3 – Readable with difficulty
4 – Readable with little difficulty
5 – Perfectly readable

*Signal Strength*

1 – Barely perceptible
2 – Very weak signal
3 – Weak signal
4 – Fair signal
5 – Fairly good signal
6 – Good signal
7 – Moderately strong signal
8 – Strong signal
9 – extremely strong signal
+10 S-meter
+20 Levels
+40 over S 9   (in decibels)

## International Phonetic Alphabet

| ALPHA   | GOLF   | LIMA     | QUEBEC | VICTOR  |
| BRAVO   | HOTEL  | MEXICO   | ROMEO  | WHISKEY |
| CHARLIE | INDIA  | NOVEMBER | SIERRA | X-RAY   |
| DELTA   | JULIET | OSCAR    | TANGO  | YANKEE  |
| ECHO    | KILO   | PAPA     | UNIFORM| ZULU    |
| FOXTROT |        |          |        |         |

There are other phonetic alphabets you can use. You can use any words to spell out a message as long as they get the message through. Many DXers use country names like N-Norway; M-Mexico; J-Japan, etc.

## Amateur Radio

In a few years, DX conditions on 27 MHz will only occur sporadically. If you really want to *talk* skip, consider getting an Amateur Radio License. Amateurs are permitted to run up to 150 watts of power with no limits or restrictions on antennas. Also, wider and less crowded bands of frequencies are available to radio amateurs (also called Hams), each one with its own unique characteristics. After all, CB is only one tiny niche in the Electromagnetic Spectrum.

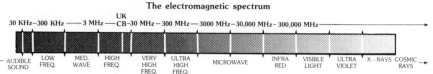

The electromagnetic spectrum

Ham radio operators can communicate via AM, SSB, FM, telegraphy, radioteletype, facsimile and even television. There are also satellites that orbit the earth and act as automatic radio relay stations, so that Hams can talk to the other side of the world, even when the skip is down. In times of natural disaster amateurs have often provided essential communications that have saved lives. And their electronic inventiveness has resulted in technological breakthroughs that have changed the course of modern communications.

Any British citizen over 14 years of age can qualify for a Ham Ticket. There are two classes of radio licenses available. the *Class A* license gives the holder full privileges on the HF DX Bands as well as all other amateur frequencies. The *Class B* license allows voice operation only on frequencies above 144 MHz. Both licenses require that you pass a multiple choice exam that tests your understanding of elementary radio theory, radio operating procedures, and UK rules and regulations. Hams are expected to know enough to operate a high-powered station safely and keep from transmitting signals that could interfere with other radio services. Home Office rules and regulations are based on international agreements made in Geneva thirty-five years ago.

Books and study materials on basic electronics and radio are available. Also, classes are sometimes held at local Ham radio clubs free of charge.

The Radio Amateur's Exam (RAE) is given twice each year, in the Spring and the Autumn. Applications must be submitted three months in advance. For more information, you can write the Home

Office for a free copy of their booklet, *How to Become a Radio Amateur*. In the Republic of Ireland, information on the Eire Amateur Radio Service may be obtained by writing the Radio Branch at the GPO in Dublin, Eire.

The Class A license also requires that you pass an exam on the sending and receiving of Morse Code. This exam is given throughout the year in London at the Post Office headquarters, at all Post Office Coast Stations, or any of the Marine Radio Surveyors Offices around the country. There are code practice records and tapes available to beginners.

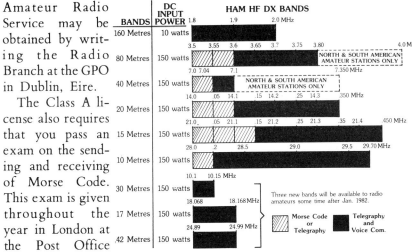

## International Morse Code

| | | | | | | | |
|---|---|---|---|---|---|---|---|
| A | di-dah | .– | Q | dah-dah-di-dah | ––.– |
| B | dah-di-di-dit | –... | R | di-dah-dit | .–. |
| C | dah-di-dah-dit | –.–. | S | di-di-dit | ... |
| D | dah-di-dit | –.. | T | dah | – |
| E | dit | . | U | di-di-dah | ..– |
| F | di-di-dah-dit | ..–. | V | di-di-di-dah | ...– |
| G | dah-dah-dit | ––. | W | di-dah-dah | .–– |
| H | di-di-di-dit | .... | X | dah-di-di-dah | –..– |
| I | di-dit | .. | Y | dah-di-dah-dah | –.–– |
| J | di-dah-dah-dah | .––– | Z | dah-dah-di-dit | ––.. |
| K | dah-di-dah | –.– | | | |
| L | di-dah-di-dit | .–.. | . | di-dah-di-dah-di-dah | .–.–.– |
| M | dah-dah | –– | ? | di-di-dah-dah-di-dit | ..––.. |
| N | dah-dit | –. | , | dah-dah-di-di-dah-dah | ––..–– |
| O | dah-dah-dah | ––– | | | |
| P | di-dah-dah-dit | .––. | error | di-di-di-di-di-dit | ...... |
| 1 | di-dah-dah-dah-dah | .–––– | 6 | dah-di-di-di-di | –.... |
| 2 | di-di-dah-dah-dah | ..––– | 7 | dah-dah-di-di-di | ––... |
| 3 | di-di-di-dah-dah | ...–– | 8 | dah-dah-dah-di-di | –––.. |
| 4 | di-di-di-di-dah | ....– | 9 | dah-dah-dah-dah-dit | ––––. |
| 5 | di-di-di-di-di | ..... | 0 | dah-dah-dah-dah-dah | ––––– |

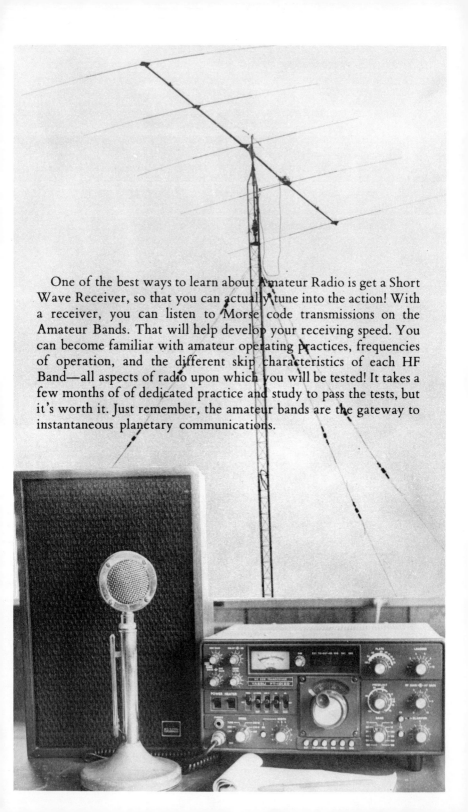

One of the best ways to learn about Amateur Radio is get a Short Wave Receiver, so that you can actually tune into the action! With a receiver, you can listen to Morse code transmissions on the Amateur Bands. That will help develop your receiving speed. You can become familiar with amateur operating practices, frequencies of operation, and the different skip characteristics of each HF Band—all aspects of radio upon which you will be tested! It takes a few months of of dedicated practice and study to pass the tests, but it's worth it. Just remember, the amateur bands are the gateway to instantaneous planetary communications.

# CHAPTER EIGHT
## Gizmos

### Power Mikes and Speech Compressors

A power mike has a built-in amplifier. Most of them have a volume control built into the case that you can adjust according to how loud you talk. Power mikes usually come as optional equipment and need to have a connector wired onto the cord to fit the radio.

Power mikes are quite popular in the U.S. and elsewhere because they can increase the modulation or loudness of almost any AM CB. AM transceivers have a built-in circuit that limits the modulation to under 100%. Most power mikes use a crystal-type micro-

phone element which has certain audio qualities that tend to "cut through" better than regular dynamic microphones. They usually run on a small battery inside the microphone, which lasts from six months to a year.

A power mike usually would not increase the deviation of an FM transceiver. The mike that comes with your FM set is selected to provide all the necessary deviation. Adding a power mike might improve the audio quality of your transmission but could result in distortion to your voice if set too high.

A speech Compressor (sometimes called a range booster) is a kind of gizmo that you plug your microphone into. Some mikes are available that even contain a compressor circuit. Speech compressors are particularly effective on SSB. They boost up your voice when you talk soft and cut it down when you are talking loud. This will raise the average power output of most SSB transceivers. If adjusted properly, they will cut down on overmodulation and pack more punch into the signal. FM CBs usually have a compressor circuit already built into the set. So the addition of this kind of mike amplifier would not make a whole lot of difference, unless you are an unusually quiet talker.

## Short Wave Receivers

A good one of these can bring you directly into contact with the entire world of High Frequency DX, including CB, Amateur, Maritime, Aeronautical and Broadcast stations. There is no better way to learn about radio than to start experiencing it directly.

There is a wide range of models available today. When purchasing a short wave receiver, there are a few things to look out for:

**Sensitivity** – good sensitivity will extend the range of your listening. Most rigs have adequate sensitivity. But any unit's ability to pull out weak signals can be improved by the addition of an outdoor aerial. An **External Antenna Jack** allows you to plug one in. Even a random length of wire stuck up in a neighbouring tree will make a tremendous difference in what you can hear.

**BFO Knob** – many of the cheaper SW sets are made to receive AM signals only. A **Beat Frequency Oscillator** mixes an extra signal into the receiver, which lets it tune into SSB and Morse Code. The duck squawk of SSB is made intelligible by fine tuning this knob. Usually it takes some practice on the part of the beginner to coordinate the dual adjustments of the main tuning knob and the BFO for adequate SSB reception.

**Digital Readout** – many of the new receivers have a numerical readout of the exact frequency you are listening to. This is a big help, especially for the beginner, when you need to locate a specific frequency. It's worth the extra quid for sure.

**Retractable Antenna and Battery Operation** – these two features allow you to take the rig with you wherever you go. Then you can squeeze in that extra hour of Morse Code practice or DX chasing wherever and whenever it becomes available.

**Stability** – this measures a receiver's ability to stay on the frequency selected. Check this by tuning in on a particular station. Listen for several minutes to see if the receiver drifts away from the station's frequency. Most receivers will display some degree of this when initially switched on. But after several minutes of warm-up, a good receiver will not drift.

Also check to make sure that when exposed to physical vibrations or shock, the frequency does not wobble or vary—an important test for any receiver with variable tuning.

**Selectable Filters** – many good rigs have filters that can be adjusted to narrow the received bandwidth to just

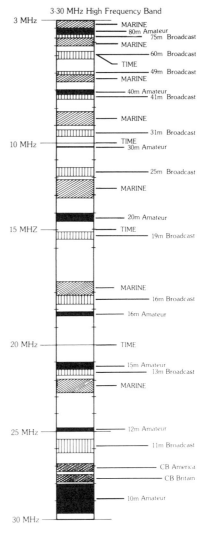

the right size, rejecting unwanted signals on nearby frequencies. This is particularly useful for picking SSB and Morse Code signals out of the QRM.

**Short Wave Bands** – some SW receivers do not cover the entire 3-30 MHz Band, while others may cover this as well as other additional broadcast bands above and below the HF frequencies. Aspiring CB DX listeners should make sure that the 26 to 30 MHz frequencies are included on any radio they buy.

## *Scanners*

A scanner is a type of receiver that can sample a number of

preselected frequencies or channels in a matter of seconds. Unlike a conventional receiver, a scanner gives instant, automatic access to these frequencies. The listener need not spend his time tuning up and down the band, looking for the action. As soon as a channel is occupied, the scanner will stop on that frequency, and the conversation will be heard. When the conversation ends, the scanner automatically resumes sampling, stopping to listen only when and where there's some action. Most scanners have a line of lights—one for each channel. They blink in succession as the receiver scans by their frequencies. All scanners have a squelch which must be properly adjusted for scanning to occur.

A scanner can cover one or more bands of frequencies for monitoring. Some of them even cover the CB channels as well as the VHF and UHF bands. Scanners lock onto the signal's carrier. That's why they are perfect for FM reception, but are not built for SSB which has no carrier and needs tuning for an understandable copy.

A scanner is capable of receiving a wide spectrum of local activities, including amateur, police, fire, medical, aeronautical, marine, governmental, utilities and business. Scanners have become very popular recently. The Home Office frowns upon the casual monitoring of most of these services as a form of entertainment. That's because many of these communications are supposed to be private and their secrecy is protected by law. But it is legal to monitor amateur radio contacts, and if you are involved in boating, flying or services that are dispatched via radio, monitoring capabilities could be an essential aid. Operators and listeners should be aware that disclosure of overheard radio telephone and other two-way conversations to other parties is illegal.

There are two basic kinds of scanners: those that are *crystal-controlled*, and those that are *programmable*. With a crystal-controlled scanner, it is necessary to purchase a crystal for every desired channel. You must know the exact frequency for each channel before ordering crystals. This limits the capabilities of the scanner to just those frequencies that you had the foresight to order However, the simplicity of these units allows manufacturers to miniaturize them. There are pocket scanners that fit into the palm of your hand and allow you to keep on listening, even while on the go.

A programmable scanner has a keyboard like a calculator that allows you to "enter in" any frequency within the scanner's range. With a programmable scanner's search feature you can sweep through whole segments of the desired band in a series of 15 kHz steps. The scanner will stop on every occupied frequency. Once you've found an interesting one, just punch it into the scanner's memory. The digital frequency display reads out every frequency that you discover. Because of these extra capabilities over a crystal scanner, the programmable scanner is also more expensive. But if versatility is what you need, you'll get it by buying one of these.

There are some other features which are usually contained in either kind of scanner. A *lockout* feature will let you temporarily remove one or more channels from the scanner's coverage. This is useful if you want to bypass a busy channel that keeps stopping the scanner. A *priority* feature lets you designate one channel so that if any action occurs there the scanner will automatically revert to it, rather than continuing to be hung up on one of the other occupied frequencies. Many scanners will run off of both 12 and 240 Volt so that either base or mobile operation is possible. Although a telescoping antenna is usually included for all scanners, the addition of an external monitoring antenna is essential for good mobile operation, and gives a much greater range to a base installation.

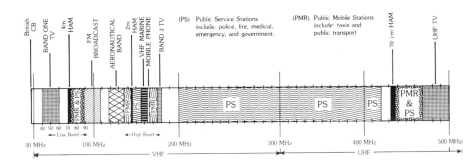

## *Walkie-Talkies*

Walkie-talkies have a variety of uses, depending on the requirements of the user and the power of the walkie-talkie itself. Some of the smaller and cheaper ones of the 100-milliwatt variety (or less) are sold as toys or for use at very close range. These are usually limited to a couple city blocks at the very most.

Because their receivers are not very selective, they will pick up just about any CB radio nearby, even if it doesn't happen to be on the same channel.

The next class of walkie-talkies is in the 250-milliwatt to 2-watt class. These are usually good for communications on a clear channel up to about a couple of miles at the most. They can be used for communication to a 4-watt mobile or base unit and will receive signals almost as well as a mobile unit. Usually this kind of walkie-talkie comes with one or two sets of crystals, and if you want any more channels than that, you have to get a set of crystals for each channel. They are usually powered by eight or ten penlight (AA) batteries. It's best to use alkaline, mercury, or rechargeable batteries, because regular batteries will run out quickly if you do a lot of transmitting.

Three to four-watt walkie-talkies are good for communications up to about five miles, maybe more, on a clear channel. There are 40 channel units available. Some of them have the kind of features you would expect in a good mobile unit: external antenna plug, P.A. earphone and microphone plugs, squelch, etc.

Almost all CB walkie-talkies come with telescoping antennas, most of them about 1.5 metres long. When using one of these, you should remember to always have the antenna extended to its full length before transmitting. This is because the SWR will be too high when the antenna is down. There are add-on antennas tht can be clamped to the stub of the telescoping antenna. These are usually short springy whips about a foot or two long with a loading coil in the base. These add-on antennas won't get out as far as the telescoping whip. They're useful for operating in close quarters where a five-foot antenna would get in the way.

If you break off one of the sections of the telescoping whip, about all you can do is replace the whole whip. This is easily done—there's just one screw holding it in at the base of the whip inside the walkie-talkie. Keep the insulating grommet in place when you put the new whip in. This will prevent it from shorting out to the case of the walkie-talkie.

For really dependable service, a four-watt handheld unit with a metal case is about the best for all-around use. If you're a dedicated CBer who doesn't want to be without ears, this might be a good thing to get.

## *Power and Modulation Meters*

These are meters that allow you to check how well your transmitter is working. A power meter reads the amount of power that your transmitter is putting out in watts (usually about 3 watts). These meters are only accurate when used with an antenna that has a very low SWR. As with an SWR meter, this meter is connected between the output of the transceiver and the antenna. A short piece of coaxial cable, with a coax plug on both ends, is used to link the metre and the transceiver. Since you are measuring the power of the carrier, a power meter will be equally accurate with AM or FM.

A modulation meter measures approximately the modulation of AM CBs only. Although there are deviation meters that can measure FM deviation, these are expensive and not readily available as a consumer item. There are even combination meters that measure SWR, power and modulation within the same package.

Some power meters also contain a field strength indicator. This requires the use of a small whip antenna which screws onto the meter. With a field strength meter you can get a relative indication of the radiated strength of a signal coming off the antenna. With the aid of this device antenna comparisons can be made and directivity of beam antennas checked.

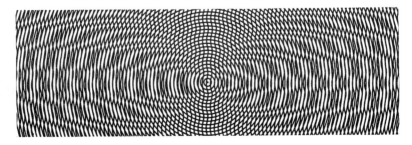

## FM and CB Receivers

There are several inexpensive combination receivers that not only allow you to receive the CB band, but also include the VHF FM Broadcast Band and the Band One and Three TV Bands. These units come with a tuning knob and a squelch control. It is even possible to pick up other stations on these including airplanes, police, fire and other public service stations. Its low battery consumption, lightweight and low cost makes this an excellent buy.

## External Speakers

Having an external speaker on your CB adds an extra dimension to listening. Better fidelity as well as louder reception can result with their use. In some instances, such as driving a big wheeler or other noisy truck, an External Speaker pointed at the driver will help out. Since most CBs end up under the dash with their speakers directed at the floor, an external speaker gets the sound up to where you can hear it.

## Antenna Switches

An antenna switch usually comes in a small box that can be mounted some place near your radio. You plug the coaxial lead-in from different antennas into the switch and run a short coax jumper to the transceiver. These are very handy when using a beam and a vertical antenna. You can monitor on the vertical and switch over to the beam to make a connection with a particular station.

## Antenna Matchers

Antenna matchers are useful when you have an antenna that has a high SWR. They can prevent the high SWR from damaging the final power amplifier of your transmitter. You'll get out much better if you use an antenna that has a low SWR than if you try to match a bad antenna system with one of these. Antenna matchers do not improve SWR at the antenna—they only make it so that the high SWR does not reach the transmitter.

## VFO

VFO stands for **Variable Frequency Oscillator**, but on the CB band they are known as sliders. They enable a CB operator to be able to slide his CB radio up above, below or in between the authorised CB channels. This device is only legal as a receive adapter. If it is used to transmit as well as receive, it is illegal.

## *Frequency Counters*

These actually read the transmitted frequency of your CB, flashing a digital readout that gives you the frequency to five or more digits. When transmitting on channel 1, for example a signal dead on frequency would read 27.60125. Frequency counters are becoming increasingly popular as base station accessories.

## *Receiver Preamps or Boosters*

These usually come in a box that plugs in between your antenna and your transceiver. They amplify all the signals coming into your receiver. These are especially good for older-model transceivers that lack volume or sensitivity on receive. Receiver amplifiers can be used on many kinds of base station radios to extend the amplification of the receiver. They operate off 240 volts AC. Some of them have a tuneable gain control, a nice feature for a base station if you don't already have it.

## *Bilateral Amplifiers*

This is a combination of a receiver preamp and a linear amplifier. In other words, it works both ways. It helps on receive and transmit.

## *Radar Detectors*

This is a piece of electronic gadgetry which detects police radar. It is sold as an accessory for mobile use. This gives you forewarning that your speed is being clocked. Some flash a light; others ring buzzers or both.

## *PA Horn*

These can be attached to any CB having a PA system. They come in either metal or plastic, (metal is more durable), and may be mounted outdoors or under your bonnet.

## *Power Reducers or Attenuators*

These are devices that hook into your coax line to reduce the effective radiated power (ERP) of your station. Some of these have a relay switch which connects the device on transmit only, so that the station's receive capabilities are not cut back along with the transmitted signal.

## Filter – (How to Stay at Peace with the Neighbours)

Folks who are into radio encounter one problem quite frequently: that is having their CB radio interfere with TV reception in the neighbourhood, causing distortion of the picture and sound when you transmit. Now, most CBs will not bother a properly built TV unless they are fairly close to one another. However, some TVs have a receiver that is not selective enough to reject a strong nearby CB signal. And occasionally a radio out of adjustment will radiate interference with its regular signal that will mess up a TV. If possible, moving the CB antenna and the TV further apart will sometimes clear it up. If this doesn't do it, you can get a low-pass filter that plugs in between your CB and your antenna. These are available at most radio supply houses. It is necessary that your antenna have a lower SWR, below 1.5 to 1 to use a low pass filter. Some low pass filters are adjustable so that you can tune them while watching a colour TV receiver. If you can tune out TV interference to your own TV, this often is sufficient proof to your neighbours that it is their TV and not your radio at fault. This can be easily demonstrated to them in your own home. If you use a tuneable low pass filter, use an SWR meter to monitor the SWR while adjusting the filter. Adjust for maximum elimination of TV interference while maintaining minimum SWR. If the interference persists, try putting a high-pass filter on the TV set. To do this, disconnect the wires connecting the TV antenna to the back of the TV set and connect them to the terminals on the high-pass filter. Then connect the leads from the high-pass filter to the TV set.

## *Dummy Loads*

A dummy load is a device that can take the place of an antenna for testing purposes. It plugs into the antenna jack on your rig and you can transmit without going out over the air. A dummy load is nothing more than a 50-ohm resistor. Small 5 watt light bulbs are sometimes used as dummy loads.

## *Tone Squelch*

Tone squelch is available as optional equipment for some transceivers. It cuts off your receiver's volume and puts it on stand by until another radio equipped with tone squelch of the same type triggers your receiver to come on. The advantage of that is that you don't have to sit and listen to other stations on a channel in order to monitor for someone else with the same tone squelch. It's a private signaling device.

## *Phaser Lasers, Gooney Birds, Roger Bleeps, and Pings*

There are several kinds of signaling devices that can be installed inside your mike or radio. They are used to attract attention. A phaser laser sounds like a sound effect from *Star Wars* and a gooney bird is like an electronic bird call. These effects are controlled by a switch so that you can turn them on anytime during a transmission. Roger Bleeps were originally used by the US astronauts. They emit a short bleep when the mike is unkeyed automatically, to let the other station know every time a transmission is finished. A ping is a tone that comes on when the mike is keyed and trails off in a second.

# CHANNEL NINE

## Emergency Procedures

There are many times when CB is the quickest or the only way to get help in an emergency situation. *You might be the one* to come across a motorist off in a ditch or in need of assistance.

First determine what kind of assistance you need and how you can best get this assistance. If you are on the superslab or motorway, there will be a lot of mobile stations on the trucker channel. Also, if mechanical assistance such as a wrecker is needed, you may find a local wrecking service monitoring the trucker channel. There may be a Smokey on this channel, too, if you need one.

It might be best to get in touch with a base station that has a land line if you can't get direct assistance.

Channel 9 is the National Emergency and Motorway Assistance channel. There are base and mobile stations all over the country that monitor Channel 9 for this purpose. Here is an example of how to get help on Channel 9:

**THIS IS \_\_\_\_(call sign)\_\_\_\_ at \_\_\_\_(exact location)\_\_\_\_ IN NEED OF ASSISTANCE. IS THERE A BASE STATION ON THE CHANNEL? OVER.**

It's likely that a station will respond to your call with his call sign and ask what he or she can do to help.

If the situation is a life and death emergency you can get on the air and break for a 10-33.

**BREAK. BREAK. BREAK.\* WE HAVE A 10-33 AT \_\_\_\_(exact location)\_\_\_\_ THIS IS \_\_\_\_(call sign)\_\_\_\_ REQUESTING \_\_\_\_(police, medical, etc.)\_\_\_\_ ASSISTANCE. OVER.**

If you can't make contact on Channel 9 try other channels. After making contact on another channel you can change to Channel 9 to pass any needed information. You should stay at the scene until you are no longer needed.

---

\*The international distress signal, **MAYDAY** may be used instead of **BREAK, BREAK, BREAK** by some marine or other stations.

If you hear an emergency situation you should immediately stop all transmitting and listen to see if the station involved is getting assistance. If another breaker is already helping out it is probably best to just listen for a while to see if it's covered.

In some situations you might have to answer the call for help. In this case, keep all transmissions as short as possible and speak clearly, giving the station your call sign and location.

Write down all information, including: station's call sign—exact location of emergency—description of emergency—type of help needed—time.

Ask the station to remain at the scene in case there is any further need for communication. The main thing is to listen closely and pay good attention.

After making any necessary phone calls, let the station know if help is on the way. Stay in contact if possible until help arrives.

It's a good idea to announce the situation on the trucker channel if it is a highway accident.

You shouldn't transmit on any of the 40 channels while you are within a mile of the emergency scene if other stations are passing emergency traffic. You might bleed over onto Channel 9 or any other channel if you are so close. It's a good idea to not use Channel 8 or 10 at all because you tend to bleed over onto Channel 9.

# CHAPTER TEN

## Fixing Your Rig
## or
## What Went Wrong

...hmmm. Looks a tad dodgy in here, wot?

If something goes wrong with your rig there are quite a few things that you can do to make it work again or to find out what's wrong with it.

One thing you have to keep in mind when working with CBs or any kind of electrical equipment is Safety First. All CB radios that run off house AC mains have high voltage in them. Definitely enough to kill you. So take it easy and pay good attention.

You can't hurt yourself with the voltages in an 12-volt mobile transistor rig but you can burn out the rig if you're not careful.

There are certain safety precautions that you need to know if you're going to take the cover off your radio:

- Never take the cover off the radio with the power cord plugged in. This includes 12-volt transistorised rigs.

- Move any metal objects or wires out from around or under the radio when testing or working on it.

- Never turn on the rig without the correct fuse in the fuse holder.

- Don't operate your rig during a lightning storm. Don't touch any components inside the radio while the power is connected.

- Don't touch or turn on the radio if it or you are wet.

Most of the time the different problems that come up with CB rigs are external to the rig itself—such as the antenna, microphone, connectors and wires.

If you are going to get inside your rig be sure you know what you're doing. Just twiddling around in an FM rig can decrease its overall efficiency.

On some occasions when your rig goes out, it will be necessary to take it to a radio repairman who has the test equipment and knowledge to do the job. But many times the problem is simple and you can figure it out on your own.

## *Trouble-Shooting Guide*

### *RECEIVING*

| Symptoms | What to Do |
|---|---|
| 1. No dial lights, no receive, set is dead. | 1. Is it plugged in? Check to see if power switch is on. Check to see if fuse is blown. If it is, replace it with another fuse of the same rating. Don't use aluminium foil. Check for possible frayed wires or skinned-off insulation. If it is a mobile, sometimes the ignition key must be on for the rig to get its juice. Check connections where the rig gets its power. Check earth connection. |
| 2. Can receive but cuts in and out. Dial lights blink. | 2. Intermittent connection to power or battery. Check fuse holder. Check ground connection. Make sure that screws are tight on all mounts. Wiggle power cord and antenna coaxial cable—see if it makes the power cut on and off. Check power plug on transceiver. |

## Receiving (continued)

| Symptoms | What to Do |
|---|---|
| 3. Dial lights come on, but no receive. | 3. Check squelch knob and make sure the mike cord is plugged in. Make sure PA switch is in CB position. Wiggle microphone cord and connector. If it cuts in and out, there might be a loose connection in the connector or mike. In this case, the mike cord might need to be cut off and resoldered. |
| 4. Receiving only very nearby stations. | 4. Is the antenna connected? How is the SWR? If the SWR is high, possibly there is a short in the coax or connectors. Check local-distant switch; should be in distant position. Check RF gain knob; it should be all the way up. |
| 5. Receiving only hiss. | 5. Check antenna and connector. Take antenna connector off and put it back again. If that makes any difference in the hiss try other channels and a radio check with a nearby station. |

## Receiving (continued)

| Symptoms | What to Do |
|---|---|
| 6. Fuse blows when rig is turned on. | 6. Short or blown transistor inside radio. Check to see if screws or mounting bolts are touching any components or wires inside the radio. Do not replace the fuse with a larger fuse or aluminium foil. This can cause further damage to radio. |
| 7. Stations received move S-meter, but do not come on through speaker clearly or not at all. | 7. Check squelch knob. If speaker sounds fuzzy, try plugging an external speaker in. If this works, you might have a blown speaker. Check wires leading to speaker inside radio. Could be bad audio transistor. |
| 8. Nearby stations sound fuzzy. | 8. If they're within 100 feet, this might be normal. Try switching local-distant switch to local, or turning RF gain down. |

*Receiving, continued*

| Symptoms | What to Do |
|---|---|
| 9. Smoke comes out of top of rig when turned on. | 9. By all means turn it off!! Pull the plug. Check to see what size fuse is in the fuse holder. It should be no more than 3 amps for a transistorised radio. Check the polarity of the battery connections in a mobile. Check to see that the negative and positive wires are hooked up right. If you are using an external speaker, make sure that none of the wires leading to it touch the chassis of the radio or the body of your vehicle. After disconnecting radio from the power, replace fuse with one of the proper value. Check to see if any screws or mounting bolts touch components inside radio. If the polarity was reversed, try out the rig again. If the fuse blows again, there is probably a blown diode or transistor. |
| 10. Receiver just hums. | 10. Microphone not plugged in. If radio uses external power supply, it might be blown. Try radio on a car battery or another DC power supply. |

## TRANSMITTING

| Symptoms | What to Do |
|---|---|
| 1. When the mike button is pushed, receiver does not cut out, cannot transmit. | 1. Check microphone cord and microphone switch. The cord might have pulled some of the wires loose in the connector. These can just be resoldered to the proper pins inside the connector. If radio has a relay, it might need to be replaced or cleaned. |
| 2. Transmits a carrier, but no modulation. | 2. The problem is probably in the microphone or cord. Check PA-CB switch. |
| 3. No carrier, no modulation. | 3. Check mike cord. Try different channels. |
| 4. Some channels not working. | 4. Could be a dirty channel selector switch. Spray switch with TV tuner cleaner. |
| 5. Fuse blows when transmitting. | 5. Blown power transistor. Screws or bolts possibly shorting against components inside radio. Short in antenna connector or antenna. |
| 6. Only getting out ¼ mile or so. Relative power meter reads zero. | 6. Check antenna and coax cable adjustment screws might need to be set. If all these are OK, final transistor might be blown. Have it replaced. |

## Receiving (continued)

| Symptoms | What to Do |
|---|---|
| 7. Transmitter "breaking up." Choppy transmission. | 7. Probably a loose connection somewhere. If it only does it on transmit, it's probably the microphone or cord. If on transmit and receive, it might be the antenna or coax cable. Check the earth on the antenna. |
| 8. Squealing on transmit. | 8. Loose wire in microphone or cord. If you are using a power mike, try turning it down slightly. If this doesn't help, try a new battery in it. Make sure that the aluminium foil shielding in the power mike is in place right if you change the battery. |
| 9. Your voice heard from speaker when transmitting. Howling. | 9. Check the PA-CB switch. Also check external speaker wires and PA speaker wires to see if they're grounding out to the body of the vehicle or radio. Possibly using low battery. Also check for shorts in microphone or connector. |
| 10. Weak modulation. | 10. Mike element might need to be replaced. Transistor blown. |
| 11. Modulation distorted. | 11. Power mike up too high. Mike gain up too high. Microphone might need to be replaced. |

# Repairing Microphone & Antenna Connections
## Tips on Soldering

All electrical soldering is done with rosin-core (not acid-core) solder. Acid-core tends to corrode electrical connections. You can test to see if the soldering iron is hot enough by touching the solder to the tip. It should melt easily. It's a good idea to practice on a couple of wires before trying to solder your microphone cable.

When soldering wires, it's a good idea to "tin" the wires before soldering. You tin the wires by heating the end of the wire up with the iron and applying a small amount of solder to the iron and the wire—just enough to put a thin coat on the wire. Make sure not to melt the plastic insulation while doing this.

Attach or stick the wire in the terminal you want to solder it to. The connector and wire should be held steadily while soldering. When you solder, put the tip of your iron so that it rests on the wire and the terminal at the same time, and apply enough solder for it to flow over the wire and terminal evenly. Never apply any more solder than is necessary to hold it good. Don't move the connection until you're sure the solder has set. It should be nice and shiny.

## Antenna Connections

The antenna is another place where a lot of connections go bad due to weathering and movement. The most common antenna problems are in the coaxial lines and connectors. An insulator at the base of an antenna can break or wear out and cause a short. Sometimes a temporary repair can be made with electrical tape.

Also, if the coax is run through an open window or door, the line can get crimped or cut. In this case, if there is enough excess coax, you can cut off the cable at the break and reconnect it to the antenna.

## Coax splices

It's best to have your coax be one long, unbroken run from your rig to the antenna. If splices have to be made in the line, you'll want to make sure there is good electrical contact made. Use two male PL 259 coax connectors and a double female connector. Generously cover this complete assembly and any other external coax connections with black electrical tape to protect them from moisture.

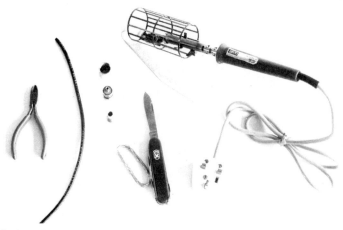

## *Soldering Coax Connectors*

You'll need some basic tools: a soldering iron, rosin core solder (60/40), medium wirecutters and a knife.

**Step One:** The most common coaxial connector on most rigs is called a PL-259. It comes apart into three separate pieces.

**Step Two:** Slip adapter and outer cylinder over the cable.

**Step Three:** Strip outer insulation back 1¼ inches. Make sure not to nick or break the copper braid underneath. Separate the strands of braided shield very carefully; don't cut or break any of them off.

**Step Four:** Slide the adapter up to the end of the black outer insulation and bend the shield down over it. Then cut the shield so that it doesn't extend down over the threads of the adapter. (The larger RG-8 coax doesn't require an adapter. Here the braided shield is folded back over the insulation.)

**Step Five:** Slip the insulation off the centre conductor of the coax, leaving a little more than ¼ inch of insulation next to the adapter. If you nick the centre wire, start over again, because a nicked wire here will eventually break if much stress if put on it.

**Step Six:** Slip the remaining part over the inner conductor and adpater, making sure none of the braid strands slip up and touch the centre conductor. (With RG 8, the remaining part of the connector slips on over the braid tightly, holding it in place without the aid of an adapter.)

**Step Seven:** Solder the inner conductor on the tip of the connector. Let the solder flow into the tip a little bit, but not too much or you'll drip it way down into the connector, shorting it out.

**Step Eight:** Cut the inner conductor off flush with the tip and screw the outer cylinder back over the rest of the connector.

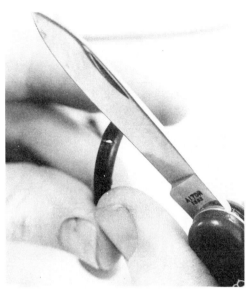

## Coaxial Connector Assembly

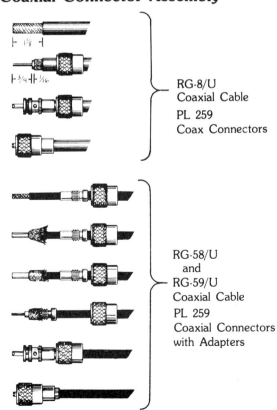

RG-8/U
Coaxial Cable
PL 259
Coax Connectors

RG-58/U
and
RG-59/U
Coaxial Cable
PL 259
Coaxial Connectors
with Adapters

## *Microphone Connections*

The microphone cord, with all its pulling and moving around, is often a source of trouble. Sometimes a wire comes loose inside the connector where it plugs into the rig. Sometimes, due to flexing, a wire will even break inside the cord where you can't see it, especially at the points where it goes into the connector or the microphone. You can usually tell when this is the case by transmitting, holding the microphone steady while talking, and wiggling the mike cord in various places.

To check the microphone connector, take it apart on a flat clean surface. For some connectors you will need a small jeweler's or Phillips head screwdriver. Take out all the screws in the connector and put them in a safe place. Then take apart the two halves of the connector, or the cover, depending on the type of connector. If one of the wires is broken off one of the pins of the connector, it's sometimes possible just to solder it back onto the pin. Make sure that it does not touch any of the other pins or bare wires.

Sometimes it may be necessary to cut the cord off and resolder the whole thing. Cut it off about an 4 cm (2") back from the connector. Then cut the outer jacket back 1 or 2 cm (1") from the end of the cable, being very careful not to nick the insulation on any of the wires inside the cable. Unwind the outer braided shield from around the inner wires and twist the small wires of the shield tightly together. Strip ⅛ inch of insulation off the tips of the inner wires and tin these tips as well as the twisted tip of the shield.

Now you're ready to solder the wire to the connector. Be sure you have slipped the outer covering or case of the connector over the cable before soldering (depends on type of connector). Now take one of the old wires off one of the pins of 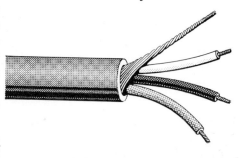 the connector and resolder the new wire of the same colour to that pin. Do all the rest of the wires the same way. Be sure you have the right colours on the right pins. The pins are usually numbered right on the connector. Make sure no stray strands of wire touch any other pins or the metal case.

Now you're ready to reassemble the connector case. Make sure the cable clamp firmly grips the outer jacket of the mike cord, taking all the strain off the connections.

For some more complicated microphones, such as power mikes, this could be more difficult, because you can't get at the connections as well. You might want to have this done by an electronics repairman.

## *Using a Continuity Tester*

Whether you are checking out old connections or double checking new ones that you've just made, a continuity tester can be quite handy. This inexpensive tool has a battery which connects to a light bulb, like a flashlight. Instead of having a switch to light the bulb the continuity tester has a thin metal probe on one end and a long wire ending with an alligator clip on the other end. When the clip and probe are touched together, the battery connection to the bulb is completed and the light goes on. The probe can be touched to one end of any wire and the clip connected to the other end. If the suspect wire is broken, the connection will be uncompleted and the bulb will not light. If the wire is unbroken (or continuous) electricity reaches the light letting you know that a proper connection exists.

This kind of tool is very useful, because you can get a visual indication of whether any particular connection is broken or shorted to earth. A mike cord can be checked for broken wires by connecting the probe end to one of the mike connector pins and the clip end to where the cable connects to the switch inside the mike. Coaxial jumpers can be checked for opens or shorts with this device. Touch the probe to one coax connector outer shield and the clip to the other connector's shield. The light should come on. You can repeat the process by connecting to the tip ends of each connector. A short will show up if you attach the probe to the tip and the clip to the outer shell. If the bulb lights in this situation you have a short.

Note: Many types of base loaded mobile and base station antennas use a shunt-type coil between the hot and earth of the antenna. This would look like a short to a continuity tester if it was hooked across the antenna's connector. If you want to check the coax, you must first disconnect the antenna from the coaxial line. Most antenna problems are either at the connector itself or at the base of the antenna where the coax hooks in.

## Noise: How to get rid of it

The object of noise suppression is to capture those loose noises made by your vehicle's electrical system and run them to earth before they have a chance to reach your receiver. These noises are usually created by sparks originating from your spark plugs, distributor, accessory motors, alternator, regulator or gauges. The sparks are electrical impulses that put out static much like little transmitters. There are various ways to keep these sparks from radiating energy that your receiver will pick up. Most of this kind of noise is picked up by the antenna. Most of the time in a car or a truck with a petrol powered engine, the ignition system (spark plugs, distributor, ignition wires and coil) causes most of the noise. Diesel-powered engines have no spark plugs, so there is no ignition noise.

Ignition noise is recognised by a loud popping or crackling, increasing to a buzz when you rev up the motor. If you rev up your engine to a high speed and then shut the key off, the moment you shut it off the ignition noise should stop.

One of the *best remedies* for this kind of noise is to install *radio resistance spark plug wires*. These are available in almost all auto parts stores. Also, *radio resistance type spark plugs* will cut down ignition noise quite a lot. After installing these the engine should be retimed and tuned at a reputable garage. If you're already using them and still get a lot of noise you might need to suppress other sources of noise.

Another source of noise is the alternator. There are various kinds of suppressors available for alternators. They're available at many places where CBs are sold. Alternator noise sounds like a whine that varies with the engine's speed. Usually, an alternator noise suppressor in the battery lead of the alternator clears up the whine.

Noise is sometimes caused by heater and wiper motors and gauges. This kind of noise can be easily remedied by the addition of a coaxial capacitor. This is the kind that's found in most radio stores—it has two screw terminals. Sometimes it might be necessary to drill a hole somewhere close to the motor or the gauge to mount it.

To hook it up you cut the hot wire leading to the motor and strip off the insulation ¼ inch back. Then connect the two ends you now have to the two screw terminals on the capacitor.

Noise from the ignition system occasionally comes through the wires leading to the radio. The noise is picked up in the engine compartment or on the battery leads and channeled to the radio by the wires. In this case, an RF choke and capacitor should be used on the hot lead going to the CB. Also coaxial cable can be run up to the battery, earthing the shield and using the inner conductor to run the juice to the rig. Make sure to use an in-line fuse where it attaches to the battery connection.

Another kind of static is caused by bare wires or loose connections shorting or arcing to the frame or other wires in the vehicle. This is usually noticed on bumpy roads. One way to find which wires are doing this is to listen on the radio with the engine stopped, ignition switch on. Wiggle wires under the dashboard and in the engine compartment. If any static is found, tape up or repair wires.

The regulator is another frequent cause of static. It's usually noticed as a clicking or intermittent popping sound. When the engine is just started up or when the headlights are turned on, it tends to come on stronger. To clear this up make sure that the regulator case and mounting screws have a good electrical connection to the body. You may also need to add .1 MFD coaxial feed-through capacitors on the battery wire leading to the regulator. Caution: Don't use this type of capacitor on the field connection of the regulator. There's a special kind of suppressor used on the field connection.

Generally, petrol-powered vehicles will be a lot noisier than diesel-powered because there are no spark plugs in a diesel engine. Spark plug or ignition noise is the main cause of static in an automobile. Even if you try all the suppression methods possible you may not be able to cut down all of the spark plug noise without making your engine not work. But you should be able to cut it down quite a lot—enough to make listening enjoyable.

For the real rough cases there are marine-type shielded ignition harnesses available. These will cut out almost all ignition noise from the spark plugs. They're fairly expensive, but they're very dependable. You would probably have to order them through a two-way communications outlet or a marine engine distributor.

Here's another thing that we've tried for cutting down static from the distributor: get a large-size tin can, big enough to just fit over the distributor. You can mount it down with a couple of angle brackets to the engine block. The can should be connected securely to the engine block to make a good earth. Make sure that your ignition wires are in good condition if you do this, because if there are any cracks in the insulation they will probably arc across to the can. You need to cut the bottom and top out of the can or make suitable holes for the ignition wires to come out of the top. Make sure you get the ignition wires back on the distributor in the proper order. There should be no sharp edges left on the can—these could wear through the ignition wires. We've experienced a drop in ignition noise by half on some vehicles using this method.

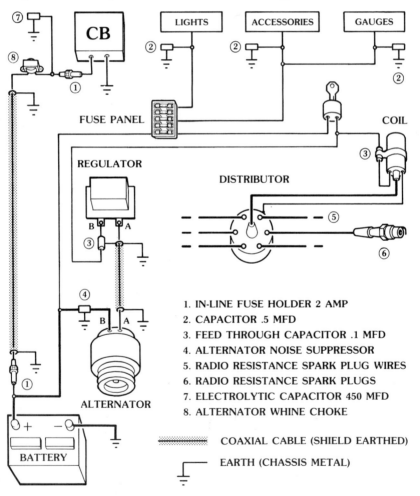

1. IN-LINE FUSE HOLDER 2 AMP
2. CAPACITOR .5 MFD
3. FEED THROUGH CAPACITOR .1 MFD
4. ALTERNATOR NOISE SUPPRESSOR
5. RADIO RESISTANCE SPARK PLUG WIRES
6. RADIO RESISTANCE SPARK PLUGS
7. ELECTROLYTIC CAPACITOR 450 MFD
8. ALTERNATOR WHINE CHOKE

COAXIAL CABLE (SHIELD EARTHED)
EARTH (CHASSIS METAL)

# CHAPTER ELEVEN
## Do-It-Yourself Antennas

There are lots of antennas out on the market today—all kinds of different shapes and sizes. Because of this competitive market, antenna companies are always trying to put out a better product for a cheaper preice. Commercially made antennas are usually easy to put up and maintain. However, you might want to try your hand at making an antenna. You can make an antenna out of readily available parts that will work as well or better than some commercially made antennas.

You'll need an SWR meter to check out the antenna after building it.

### ¼ Wave Ground Plane Antenna

This antenna consists of a driven element and four radial wires which act as a ground. The driven element receives the transmit energy from the rig.

**Parts List**

| | |
|---|---|
| 100″ (254 centimetres) piece of aluminium pipe or conduit, ¾″ or 1″ diam. | Rope, enough to guy the earth radials, depending on the height of the antenna |
| Two U-bolts, same size as pipe | A couple of two by fours |
| One sheet metal screw | Electrical tape |
| One J-hook | Silicone seal to cover coax connection |
| Four "egg" ceramic insulators | |
| 400″ (10.2 metres) of 16 gauge wire | |

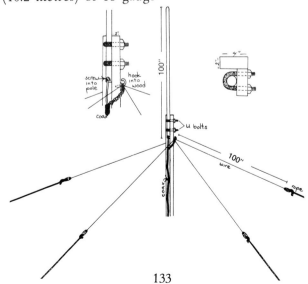

The inside conductor of the coax is connected to the aluminium pipe by means of a screw into the bottom of the pipe. See the detail drawing on preceding page. Coat this connection with sealer or cover it with tape to protect it from corrosion.

All vertical antennas need to be earthed in some way. A mobile antenna uses the car body as the earth. On this antenna, the four radial wires are used as the earth. This is called the ground plane of the antenna.

The braided wire which forms the outside conductor of the coax is soldered to all four radial wires. The wires must be exactly 254 cm (100″) long (¼ wavelength).

Remember that the inner conductor and outer braid of the coax must not touch each other, and the radials must not come in contact with the driven element. The radials slope down at about a 45° angle in different directions, and are tied to the insulators. Rope or nylon cord is then tied to the insulators and used to hold the radials out. They can be attached to anywhere convenient—trees, fence, house, etc.

If you are short on room for such a radial system, you can use 254 cm (100″) pieces of aluminium tubing, or suspend the wires on PVC pipe bamboo, or 1″ x 2″ wooden sticks.

It's a good idea to check the SWR when done. It should be lower than 2, and most likely lower than 1.5 or 1.3.

A ¼ wave ground plane made from wire can be suspended from a tree. We've talked to stations over 40 miles away using this antenna up about 30 feet high, running a mobile rig for a base.

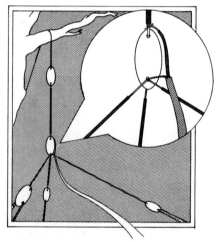

## Using a Mobile Antenna for a Base Station

A mobile antenna can be used as a base antenna by mounting it on the top of a metal pipe. The metal pipe serves as earth connection for the antenna, taking the place of the body of a vehicle. Remember to run a separate earth wire to a proper earthing rod for lightning protection.

## Building a Gain Vertical

This is an easy antenna to build and find the parts for. If you follow these instructions and have it come out looking like these pictures it should have a very low SWR and appreciable gain.

You can use any number of pieces of aluminium pipe so long as they are ridged and fairly thick walled so as not to get bent and broken in a strong wind. The pieces should be gradually smaller, one being able to fit inside the next. Cut two 2" grooves with a hacksaw down the outer pieces of aluminium and put a hose clamp around them. Now adjust the antenna to 22½ feet (6.85 metres) and tighten the hose clamps down to hold it all together. This 22½ foot vertical element can now be mounted with two U-bolts on to the 2" x 6" board. This board should be treated or painted to protect it from the weather.

Put a bolt through the piece of wood a few inches below the vertical element. Here you should fasten one end of the 6 foot (183 centimetres) piece of copper wire or tubing, the outside braid of the coax and each of the 100 inch (254 centimetres) long, stranded wires. These are called the ground radials and should be tied off with string (not wire) at a 45° angle away from the bolt. The

135

ground radials and the braid from the coax can be soldered together or can be crimped together with a crimp connector which fits the bolt. The other end of the 6' copper wire is bent and fastened to the vertical element as the picture illustrates. The end of the centre wire of the coax is then twisted onto this in such a way that it can be slid up or down along the copper wire and soldered after adjustment.

The SWR of this antenna is adjusted by sliding this connection. You do this by keying your rig up on channel 20 and sliding this connection up or down until you have the lowest SWR. In our experience, an SWR of 1.1 to 1 was easily reached on channel 20 with a low SWR throughout the 27 MHz band.

Be sure to tape up the end of the coax real well so no moisture gets in it.

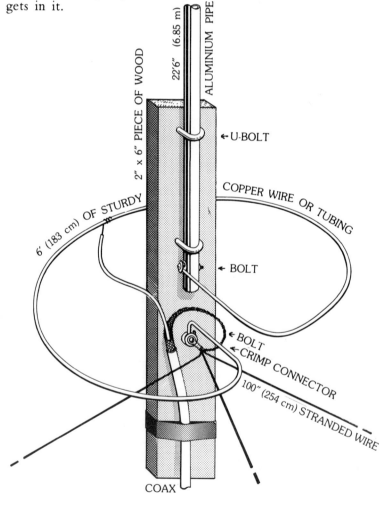

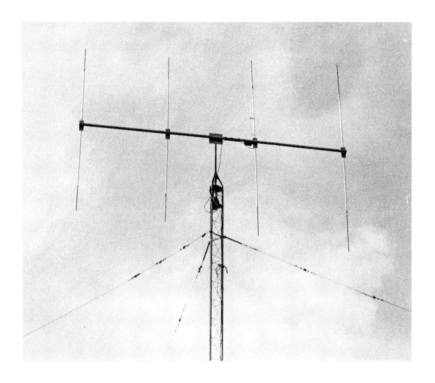

## *The Long John Antenna—Build a 10 dB gain beam!*

A 4 element Long John antenna will make your radio *ten times more powerful* than if you used a ¼ wave vertical antenna. This type of high gain beam is easy to build. Materials include aluminium plates and tubing, automotive muffler clamps, and radiator hose clamps—all of which are easy enough to find in any large city.

### Parts List

4 sections of 1" (2.5 cm) aluminium tubing, 12 feet (3.7 m) long with .035" (1 mm) wall
3 sections of ⅞" (2.2 cm) aluminium tubing, 12 feet (3.7 m) long with .035" (1 mm) wall
1 section of 2¼" (5.6 cm) OD (outside diameter) aluminium pipe 20 feet (6.1 m) long with a thick wall size
1 section ½" (1.25 cm) aluminium tubing 2 feet (.6 m) long, any wall thickness
12 2¼" (5.6 cm) muffler clamps with lock washers
8 1" (2.5 cm) radiator hose clamps
2 2" (5 cm) muffler clamps
8 Sheet metal screws—No. 8 self tapping
1 plastic box 4" x 8" x 2 " (10 x 20 x 5 cm) with sealable lid

4 aluminium plates 4" x 12" x ⅛" (10 x 30 x .3 cm) thick
1 aluminium plate 4" x 10" x ⅛" (10 x 25 x .3 cm) thick
1 aluminium plate 12" x 12" x ⅛" (30 x 30 x .3 cm) thick
1 coax connector-female chassis mount
1 tuneable capacitor 0-100 pF-receiver type. (This kind of capacitor is used in many receivers as the frequency control, and can be bought or retrieved out of an old MW radio or other radio junk.)
12 bolts ⅛" (.3 cm) thick x 1" (2.5 cm) long with extra lock washers and nuts
1 plastic knob to fit the capacitor's shaft size
8 U bolts 1" (2.5 cm)
1 piece of aluminium sheeting 12" x 12" x 1/32" (30 cm x 30 cm x 1 mm) thick
1 small piece of plexiglass 1" x 3" x ⅛" (2.5 x 7.5 x .3 cm) thick
1 tube silicone sealer

**Tools Needed**

hack saw
adjustable spanner
screw driver
pocket knife
electric drill

wire cutters
sheet metal shears
metal file
tape measure
assorted drill bits

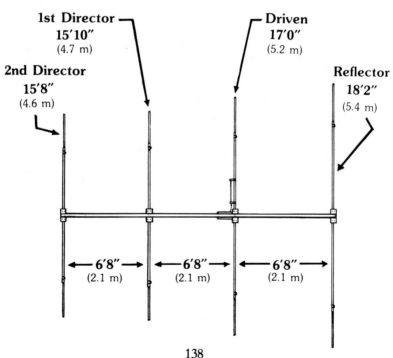

This antenna was designed to have a wide bandwidth so that it could tune from 27.000 to 28.000 MHz with a low SWR. The 6'8" (2 m) spacing between the elements and the 1" (2.5 cm) element tubing size of the elements themselves both contribute to the wide bandwidth. It is possible to build a 5 element beam on the same length of boom using closer spacing (as low as 5 feet or 1.5 m) between the elements—but only at the expense of the antenna's bandwidth. Close spacing will also result in less overall gain. In fact a close spaced 5 element beam will have no more gain than a wide spaced 4 element one! Close spacing of the elements also makes the antenna tuning adjustment more critical. If the large size of a 4 element beam puts you off, you may want to eliminate the second director element and have a 3 element beam on a shorter boom But keep the wide spacing between the elements for best overall performance.

## *Buying The Aluminium*

If you can't get the exact wall thickness of tubing that we list, get as close as you can. Make sure that the smaller ⅞" tubing will still fit inside of the larger tubing. If the fit is loose, you may have to make some shims out of sheet metal stock to take up any slack. If you buy your aluminium plates from a sheet metal company, you can often get them to cut the pieces to the exact size you need. We recommend this over having to cut all the pieces out yourself by hand.

It is possible to use a slightly smaller or larger diameter pipe for the supporting boom. But if you switch to another size, you'll also need to change the size of all boom associated muffler clamps.

## *Assembly of the Beam Antenna*

Drilling the aluminium plates – use a drill bit big enough to allow the muffler clamps to comfortably slide through the holes. The 4" x 12" plate will be used to mount the elements to the boom, and the 12" x 12" plate is for mounting the beam to a supporting pole or mast.

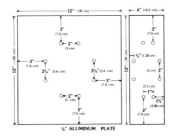

**Making the elements** – slice a notch about 4 inches into the ends of each 1″ aluminium tube. Make sure that the notch is centred in the middle of the tubing for its full length. File away any aluminium burrs that occur around the notch.

Next cut the ⅞″ tubing into two short pieces for each element. These are the tips of the elements and their length will differ depending on whether they are for a reflector, driven or director element. Here are their dimensions:

2 pieces - 4 feet long (1.2 m)    – reflector tips
2 pieces - 3½ feet long (1.1 m)   – driven elements
2 pieces - 3 feet long (.9 m)     – first director tips
2 pieces - 2 feet 10″ long (.85 m) – second director tips

Put a mark on each tip piece 1 foot from an end. Slip the reflector tips into each end of a 1″ tube so that 1 foot of the tip is inside the larger 1″ pipe and the rest extends outside of it. Repeat this for the directors and driven elements.

Slip a hose clamp over each junction of 1″ to ⅞″ pipe. Place the clamp over the 4″ slit and tighten until the two sections of pipe are firmly held together. Drill a hole sightly smaller than the size of a No. 8 self tapping screw about 8 inches back from each end of the 1″ pipe. Putting sheet metal screws in helps to anchor the tips to the 1″ centre sections of the elements.

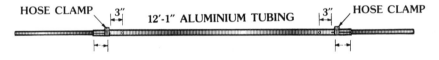

**Mounting the elements to the boom** – centre each element on one of the 4″ x 12″ plates. Put the 1″ U bolts on and tighten them into place so that the tubing is securely held, but not dented or crimped.

Mark off the 20-foot boom into 6′8″ lengths. Centre the plate mounted elements on each of these marks with the reflector first, the driver second, and then the first and second directors. Bolt them in place with the 2″ muffler clamps. Make sure that the elements all line up together.

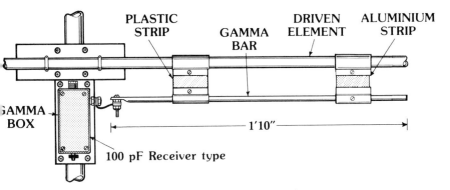

## Feeding the Antenna—The Gamma Match

A beam antenna like the one we are making is fussy about how the radio's power is delivered to it. The antenna must be fed through a balancing device which *matches* the antenna to the output of the radio and coax. Without the addition of the gamma match, the antenna would have a high SWR and much of the radio's power would be reflected away from the antenna back into the coax. With a gamma match, the antenna can be tuned for a low SWR and the maximum transfer of power from the radio on out to the antenna takes place.

**Making the gamma match** – the match connects a tuneable capacitor in line between the coaxial cable and the driver. This capacitor is mounted inside a watertight plastic container to protect it from the weather. Something like a sealable plastic sandwich box could even work. The coax connector mounts on one end of the box and let's you plug the cable in. A bolt mounted through one side of the box connects to the driven element through a short section of ½" tubing, called the gamma rod. There is a short jumper wire between the bolt and the gamma rod.

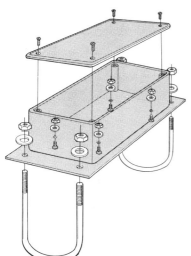

Mount the female coax connector by drilling a good sized hole centred in one end of the plastic box. Carefully widen this hole with your pocket knife until the round part of the connector will fit snugly in place. Mark the four holes that go through the connector and drill them out. Use small bolts to secure this connector to the side of the box.

In the other end of the box drill a hole just big enough to allow the capacitor's tuning shaft to snugly fit through. Mark onto the plastic box where the capacitor's mounting holes should go, and drill them out. Bolt the capacitor in place.

Centre the 4" x 8" plastic box on top of the 4" x 10" aluminium plate. Mark the areas on either end that extends beyond the box. Within that extra space drill holes for mounting a 2¼" muffler clamp on each end of the plate. Take the plastic box and drill an ⅛" hole in each corner. Centre the box again onto the plate and mark the holes onto it. Drill them out and mount the box onto the plate.

Drill a hole in one side of the box and connect a bolt to it. Add on the nuts and washers in the order pictured below.

The tuneable capacitor consists of two sets of metal plates that can be meshed together by turning the shaft. Solder a wire to the inner conductor of the coax connector. Attach the other end of the wire to the mounting terminal on the back of the capacitor that connects to the fixed metal plates. Solder this wire to

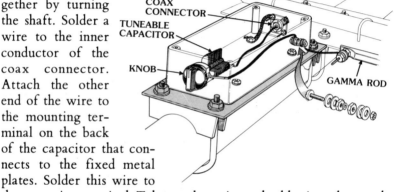

the mounting terminal. Take another wire and solder it to the metal tab that connects to the moveable plates attached to the capacitor's shaft. Connect the other end of this wire to the bolt that goes through the side of the plastic box.

Take a short piece of copper wire and use it as a jumper between the coax connector's metal base and the closest bolt that fastens the box onto the aluminium plate. Make sure that this jumper is well secured and that it cannot short out to the inner conductor wire that goes from the connector to the capacitor.

Apply some silicone sealer around the rim of the top and anywhere else where you might think that water could get in. Fasten the lid onto the box. Position the gamma box on top of the boom right next to the driven element, and bolt it in place with a couple of muffler clamps.

*The Gamma Rod* – Take the section of ½" aluminium tubing and flatten one of the ends. Drill a hole in the flattened end and put a bolt through with a nut and washers.

There are two clamps which fasten the gamma rod to the driven element. The all metal clamp is made by bending a 2" x 8" strip of aluminium sheeting around the two tubes leaving a spacing of about 3" between them. The second clamp is made out of two short pieces of aluminium with a plexiglass insulator in between the tubes. Unlike the first clamp which is the electrical connection between the rod and the driven element, this second clamp is merely there to give added physical support to the gamma rod.

Run a short jumper wire between the bolt protruding from the gamma box and the bolt on the end of the gamma rod.

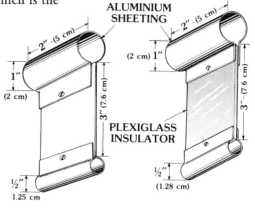

*Tuning the Gamma Match* – in order to do this, you'll need the help of a friend or two. Tilt the antenna back on its reflector and point it straight up in the air. Make sure that no stress is put on the reflector element itself, but on the boom pipe and element plate only. Hook a SWR meter up to the gamma box with a short coaxial jumper. Hook the coax into the other end. Have someone key up the CB on channel 20, which is in the middle of the CB Band. Don't worry, the two watts or so that your radio puts out will not hurt you. Calibrate the meter in the forward position and put it into the reflected position and take a SWR reading. Adjust the gamma capacitor for the lowest SWR reading. If you can't get the SWR below 1.5 to 1, stop transmitting and adjust the metal clamp that connects the gamma rod to the boom. Move it toward the far end of the gamma rod and take another SWR reading. If the SWR goes down, then readjust the gamma capacitor for the lowest reading. If you still don't have below 1.5 to 1, move the clamp a little further down the rod, until you can get a low SWR. If the SWR went up when you moved the clamp, you may have to reverse directions and move closer to the boom. By adjusting the capacitor and varying the location of the metal clamp connecting between the gamma rod and driven element, you should be able to get a SWR below 1.5. You may have to fine tune your SWR once it is in the air but it should be pretty close to what it was on the ground. Once you have the beam all tuned up and hoisted up there give someone a shout.

# CHAPTER TWELVE

## CB Radio in Europe

During the last few years CB has been growing in many European countries, including Austria, Germany, Norway, Switzerland, Belgium, Italy, Greece, Cyprus, Luxembourg, Spain, San Marino, Portugal and Yugoslavia. The governmental regulations concerning CB radio vary from country to country. The Congress of European Post and Telecommunications (CEPT) is attempting to get all European countries on a standardised 27 MHz FM system. But to this date they've had little success. In most countries of the world there is AM and SSB CB activity on the 27 MHz American channels. The following are some of the countries that have an active CB scene:

### EIRE

CB radio has become very active in Ireland over the last three years. Because of Ireland's lack of laws concerning the use of personal radio communications, the Irish customs and excise officials have allowed the unrestricted importation of CB radios into the country. CB was considered neither legal or illegal, and so far over 6 million pounds of customs duties have been collected by Irish customs officials. In fact, many of the Irish police (Guarda) already have CB sets in their vehicles, and consider them to be very useful in carrying out their duties.

Now the Irish government has proposed the legalisation of an Irish CB service with 40 channels on 27 MHz. The new service will use FM and the radios will put out 4 watts. The estimated 100,000 Irish CBers who already have AM and SSB equipment will be given a two year amnesty period. At the end of this time, all AM and SSB CB radios will be considered illegal to operate, and only the government approved FM equipment will legally be available.

## *France*

The French government legalised a CB radio service in December, 1980 in the wake of widespread illegal activity throughout the country. The French system has 22 channels with 2 watts of FM transmitted power allowed on 27 MHz.

## *Holland*

The Dutch PTT has allowed the use of CB radio for several years now. Initially, American CB equipment was allowed into the country, but now that equipment is illegal for use there. The Dutch now have a 22 channel FM CB service on 27 MHz. Because Holland is such a small country, with over 12 million people living close to one another, the government has initially restricted the transmitted power allowed to ½ watt. But due to to CB's wide popularity within the country, there is an increased demand for a wider communications coverage and more room for Dutch CBers now sharing the existing overcrowded channels. So the PTT has committed itself toward an eventual expansion of CB services in Holland to 40 channels and 2 watts of power on 27 MHz.

## *Italy*

Widespread operation of AM and SSB in Italy is heard regularly all around the world. The Italians are well known for their powerful signals. In Italy, the legal supervision of the radio airwaves is non-existent, so just about anything goes. In Rome alone there are over a dozen unauthorised television stations operating on a regular basis. On the CB band many stations run a kilowatt of power or more!

## *Scandinavia*

CB radio is very popular throughout these countries. It is estimated that in Denmark alone there are over 300,000 CB breakers. CBers there can use 23 channels with FM or AM modulation allowed on 27 MHz. Many ships around the Danish coast use CB radios and channel 11 is used as a CB marine breaking channel. Finland has 22 channels with ½ watt transmitted power allowed.

# CHAPTER THIRTEEN

## How Far Can My CB Radio Talk?
## (The CB Fanatic's Guide To Getting Out)

This is the most asked question from CB radio operators around the world. Obviously the answer to this question depends on a number of factors. What follows are ten questions that you should ask yourself concerning your present CB radio system. Each question concerns one of the basic factors of CB communications.

1. **How much power does your CB radio put out?**

   The more power that your CB radio puts out, the farther away you can be heard.

2. **How high and clear of obstacles is your antenna?**

   The higher that the antenna is and the clearer shot that you have toward the desired direction of communication, the better. This makes a definite difference in both the transmission and reception of signals. Mobiles get out best from a higher elevation.

3. **What kind of mike are you using?**

   On AM CB rigs, a power mike can increase your modulation. The closer your modulation is to 100%, the better. On FM CB radios a power mike might help, but could make your signal sound worse. On SSB rigs, a speech compressor microphone can increase the amount of average power that your radio is transmitting.

4. **How is your SWR?**

   The closer that your antenna is to a 1:1 SWR, the better. This is an indication of how well your radio and coaxial line is matched to your antenna. A good match indicates an efficient use of the power transmitted.

5. **How much gain does your antenna have?**

   A gain antenna makes more effective use of the transmitted signal, shaping and directing it for the most optimum performance. The higher the gain of an antenna, the more *Effective Radiated Power* produced from your station.

6. **How much of your transmitted power is reaching the antenna?**

   Losses in the coax can dissipate some of your signal. To cut down on this loss have your radio as close to the antenna as is practical. The thicker RG 8 coax has less line loss than the skinnier RG 58. Each connector in your coaxial line will cause you to lose as much as 1 dB of signal. This is true for any meter, coaxial switch or other add-on device: any interruption in the coax will cause you to lose a little.

### 7. What kind of antenna polarisation are you using?

For local communications, vertical polarisation generally works best. DX communications can utilize either polarisation, with horizontal sometimes offering the advantage of limiting interference from local stations.

### 8. What mode of communication are you using?

On SSB, your range of communication will be more than twice as effective per watt of power than an AM signal would be. With FM, local communications are superior to AM signals because of increased clarity and less susceptibility to noise and static. So far there are few DX stations operating FM—a definite limiting factor.

### 9. What time is it?

The time of day, of the year and of the 11 year sunspot cycle all directly affect our ability to communicate. Local contacts are often severely limited by high level signals skipping in from other parts of the world. For best local communications, it is essential to select the correct time, when skip conditions are at a minimum. DX communications, on the other hand, can only take place during the times when the atmospheric conditions are favorable.

### 10. What channel are you on?

Obviously, the clearer the channel that you are on, the further that you can communicate. This is equally true for both local and DX communications. Also, if you are operating SSB or FM it is important to be using channels that others operating these modes will be listening to. When skip conditions are good, it is difficult to locate a clear space anywhere within the American 40 channels, because so many stations around the world are operating simultaneously. That is why CBers in many parts of the world now operate in broad sections of frequencies above and below the American CB band. They are looking for a little peace and quiet. The British FM CB channels are in the area above the regular AM channels.

The better that your station is, the more opportunities that you will have for maximum range of communications. So it's really up to you—the sky's the limit!

# CHAPTER FOURTEEN

Break 14 for a copy! hilk! hilk

## CHANNEL JIVE (CBers Lingo)

We collected this list of CB jargon while listening to breakers. We noticed that a lot of American lingo is used as well, so we have included a list of the more common American slang terms.

A bit 10-1 - Weak or fading out
Ace - CBer with an unjustified high opinion of himself
Alligator Station - "All mouth and no ears"
Ally Pally, Alexander's Palace - A favourite London DX point for mobiles
Appliance operator - A CBer who doesn't know anything about his radio
Aggro - Aggravation
Ancient Modulator - Station using Amplitude Modulation (AM)
Angle Modulation - an obscure term for Frequency Modulation not to be confused with AM
Auntie Beep - The BBC
Back Door - Vehicle behind you watching what's on the motorway
Back side, Back stroke, Bounce around, flip-flop - Return trip
Back off the hammer - Slow down
Bad scene - Crowded channels
Bail out - Leaving the super slab at a roundabout
Band Bender - SSB operator
Bean store - Restaurant or road stop where food is served
Bear - Police
Bear bait - Someone driving fast without a radio
Bear den - Police station
Bear's lair - Police station
Bear bite - Speeding ticket
Barefoot - Transmitting without a burner
Beaver - Small, furry, large-toothed animal that lives in the water
Bells - Hours; 9 bells is nine o'clock
Bending my ears - Putting me on

Bending my windows - Real good copy
Better half - Your wife or husband
Between the sheets - To sleep
Big brother - Home Office, GPO
Big Circle - North Circular Rd. in London
Big Dummy - Truck driver term
Big Wheels - Lorry
Big slab - motorway
Big Smoke - London
Big Twig - 9 foot whip
Bird cage - Heathrow Airport
Black box - CB radio
Bleeper breaker - Transmitting a bleep when you let off your mike, signifies end of transmission
Blue light - police vehicle
Boat anchor - A big old radio
Bottle shop - Pub
Bottom of the shop - Channel 1
Break - Call a station
Break, Break; Breakity, break; Breaker Break, etc. - What you say to get on a channel
Breaker - CB user
Breaker on the side - Station trying to break an ongoing conversation
Breaking up - Your signal is cutting on and off
Bring it back - said at end of transmission
Brown Bottles - Beer
Brown bottle Shop - Pub
Burner - Radio power amplifier
Buzby - GPO
Candles - Years
Chalk a block - All channels filled up with breakers
Channel master - someone who tries to monopolize or control the activities on a channel
Chicken coop - Truck inspection station
Coming out the windows - Real good copy
Convoy - A few vehicles trucking together
Cop Shop - Police station
Copy? - Did you get that last transmission?
Copy, copy! - Got a copy on you!
Copying the mail - Listening to folks talk on the channel
Cubs - Police
C'mon - Go ahead and transmit
Dinosaur juice - Petrol
Dodgy - Risky
Dog biscuits - dB, decibels
Don't feed the bears - Don't get busted
Double L - Land line or telephone
Doughnut - Roundabout
Draggin' wagon - Wrecker
Dusting your knickers - Keying down on top of another station
DX - Long distance, skip

Ears - Antennas
Ear wigging - Monitoring, listening on the side to an ongoing conversation
Easy chair - Sitting in the middle of a convoy, also called the rocking chair
Eights, Eighty-eights - Hugs and kisses, good wishes, good numbers
Eleven Metres - CB band

Eyeties - Italian stations
Factory - Place of work, any kind of place for work
Ferry Lights - Traffic lights
Fetch it back - Same as bring it back
Feed the bears - Pay a ticket

Final - Last transmission; also, final power amplifier circuit
First person - Your name
Flappers - Antennas, twigs, etc.
Flat side - Horizontal polarisation (also - going to sleep)
Flat talking - Talking on the ground wave
Fluff stuff - Snow
Front door - Vehicle ahead of you, or at the head of a convoy
For fer sure - Definatory, 10-4
Full quieting - When a station is coming through with no background noise
Gear Jammer - Truck driver
Go juice - Petrol
Going breaker break - Leaving the air
Going down - Signing off the air
Golden numbers - Good numbers
Good lady - Female CBer
Good numbers, goodly numbers - Best regards and good wishes
Green light - Clear road on up ahead
Green shield stamps - money
Ground clouds - Fog
Handle - CB nickname
Ham - An amateur radio operator
Hammer - Acclerator pedal
Harry Rags - Cigarettes
High channels - Channels above 40
High numbers - Same as good numbers
Hit the sheets - Going to sleep
Holding on to your mudflaps - driving right behind you

Home twenty - Location of your home
Honey bear - Female peace officer
Horizontal, flat side - Go to bed; also horizontally polarised antenna
Idiot box - TV
In a short - Soon
In a short short - Real soon
Is that a four? - 10-4?
Jam jar, jam butty - Same as jam sandwich
Jam sandwich - Police vehicle white with red stripe through the middle
Jaw jacking - Long winded conversation
Keep it clean and don't be seen - Don't get busted
Klicks - Kilometres
Knock it down - Go down to another channel
Knock it up - Go up to another channel as in "knock it up one."
Keep 'em between the ditches - Have a safe trip
Keep the shiny side up and the greasy side down - Have a safe trip
Land Line - Telephone
Lay-by - Lorry pull-off
Linear - Linear amplifier, illegal amplifier of signal, burner
Little wheels - Car or other four wheeled vehicle
Looney-channels - Channels where the ripoffs hang out
Lucky number - A CB channel (hopefully clear)
May Day - International emergency distress call
May the blue light never shine on you - Hope you don't get busted
Marker - click marks along the motorway
Messers - Stations that deliberately cause interference
Modulation is 100% - Your signal sounds real good
Motion lotion - Petrol
Mud - Coffee
Muppet show - AM 40 channels
Negatory - No
Nelly Kelly - TV
Nerd - silly CB user
Nice one - Good copy
Noddies - Motorcycle police
Non-sus twig - A disguised CB antenna
Nosebag - A meal
On channel - On the air
On frequency - In calibration
On the peg - Legal speed limit
On the side - Standing by on the channel
One armed bandit - Petrol pump
One eyed monster - TV set
Over your shoulder - Behind you
Over - End of transmission, your return
Pedal to the metal, hammer down - Accelerator to
  the floor
Peeling off - Getting off the motorway
Personal - Your name
Pick a lucky number - Pick a CB channel to move to
Plain wrapper - An unmarked police car
Post - Marker

Pounds - Watts, notches on the "S" meter
 "Your're putting about 9 pounds on my meter."
Pull the big switch - Turn off the CB, go off the air
Pulling cobwebs over me eyes - Tricking me
Pushing big wheels - Driving a lorry
Pushing candles - How many years old
Pushing wheels - Driving a vehicle
Puts me off - Makes me uptight
Ratchet jaw - Someone with a lot to say
Ratcheting - Talking
Reading the mail - Listening to the channel
Rig - CB radio; also big truck or vehicle
Riot squad - Neighbours complaining about TV interference

Roger Bleep - A CB signaling tone device
Roller skate - A small car, such as a compact or import
Round about - Intersection on the super slab
Rubber bander - Green CB operator, beginner on CB
Russian Woodpecker - Russian long range radar that sounds like
 a woodpecker and causes interference to CB
Seat cover - Female passenger
Sidewinder - SSB operator
Shoes - Burner

Slider - VFO, a device used to make CB's tune all across the channels, in between them, and then some
Smokey, smokey the bear - Police
Smokey report - Location of smokey the bear
Smokey on rubber - Bear on the move
Smokey town - London
Spaghetti junction - Birmingham
Sparks - Radio technicians
Squeakee, squeaker - Rip off station, usually talks in a high squeaky voice
Square wheels - A stopped vehicle
Station on the side - CBer wanting to join on-going conversation on the channel
Stateside skip - Skip from the USA
Stinger - Top part of a CB antenna usually a stainless steel whip
Suicide jockey - Lorry driver carrying hazardous materials

Super slab - Motorway
Taking pictures - Police using radar
Ten-ten till we do it again - Sign off
Ten-um-ah-four - 10-4
That's a fox - 10-4
Top of the shop - Channel 40
Tower Town - Blackpool
Treading wheels - Same as pushing wheels

Trip around the horn - Scanning through the channels
Twig - Antenna
Twins - Dual antennas
Vertical - Vertical ground plane antenna
Vertical side - Vertical polarisation
Walking all over you - Another louder station is covering up your signal
Wallpaper - QSL cards, exchanged by CBers, that have their call sign, handle and location printed on them
Wall to wall and tree top tall - Loud and clear signal
Wall to wall spaghetti - Overcrowded with Italian skip
Wally - Crazy operator
Watering hole - Pub
We up, we down, we gone bye bye! - Sign off
Windup - A put on
Wrapper - Vehicle
XYL - Wife
YL - Young lady
Z's - Sleep
Zoo - Police headquarters

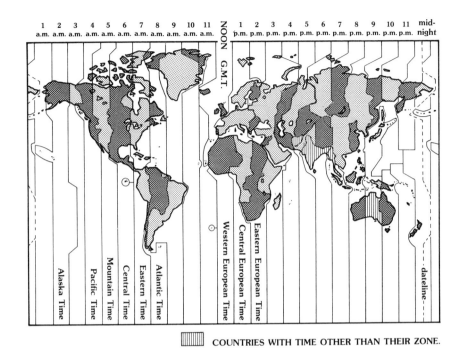

## Time and Metric Conversions

| | | | | | |
|---|---|---|---|---|---|
| 1 millimetre | (mm) = | .039 | inches | | |
| 1 centimetre | (cm) = | .39 | inches | | |
| 1 metre | (m) = | 39.40 | inches | 3.28 | feet |
| 1 kilometre | (km) = | .62 | miles | | |
| 1 inch | (in.) | 2.54 | cm | | |
| 1 foot | (ft.) | 30.48 | cm | | |
| 1 yard | (yd.) | .914 | metre | | |
| 1 mile | | = 1.6 | km | | |

## TEN CODE Used by CBers

| | | | |
|---|---|---|---|
| 10-1 | Receiving poorly | 10-44 | I have a message for you (or for ___) |
| 10-2 | Receiving well | | |
| 10-3 | Stop transmitting | 10-45 | All units within range please report |
| 10-4 | OK, message received | | |
| 10-5 | Relay message | 10-46 | Assist motorist |
| 10-6 | Busy, stand by | 10-50 | Break channel |
| 10-7 | Out of service, leaving air, not working | 10-55 | Intoxicated driver |
| | | 10-60 | What is next message number? |
| 10-8 | In service, subject to call, working well | 10-62 | Unable to copy, use phone |
| | | 10-63 | Network is directed to ___ |
| 10-9 | Repeat message | 10-64 | Network is clear |
| 10-10 | Transmission completed, standing by | 10-65 | Awaiting your next message |
| | | 10-66 | Cancel message |
| 10-11 | Talking too fast | 10-67 | All units comply |
| 10-12 | Visitors present | 10-68 | Repeat message |
| 10-13 | Advise weather/road conditions | 10-69 | Message received |
| | | 10-70 | Fire at ___ |
| 10-16 | Make pickup at ___ | 10-71 | Proceed with transmission in sequence |
| 10-17 | Urgent business | | |
| 10-18 | Anything for us? | 10-73 | Speed trap at ___ |
| 10-19 | Nothing for you, return to base | 10-74 | Negative |
| | | 10-75 | You are causing interference |
| 10-20 | My location is ___ | 10-77 | Negative contact |
| 10-21 | Call by telephone | 10-81 | Reserve hotel room for ___ |
| 10-22 | Report in person to ___ | 10-82 | Reserve room for ___ |
| 10-23 | Stand by | 10-84 | My telephone number is ___ |
| 10-24 | Completed last assignment | 10-85 | My address is ___ |
| 10-25 | Can you contact | 10-88 | Advise phone number of ___ |
| 10-26 | Disregard last information | 10-89 | Radio repairman needed at ___ |
| 10-27 | I am moving to Channel ___ | 10-90 | I have TV interference |
| 10-28 | Identify your station | 10-91 | Talk closer to mike |
| 10-29 | Time is up for contact | 10-92 | Your transmitter is out of adjustment |
| 10-30 | Does not conform to rules | | |
| 10-32 | I will give you a radio check | 10-93 | Check my frequency on this channel |
| 10-33 | EMERGENCY TRAFFIC AT THIS STATION | | |
| | | 10-94 | Please give me a long count |
| 10-34 | TROUBLE AT THIS STATION HELP NEEDED | 10-95 | Transmit dead carrier for 5 seconds |
| 10-35 | Confidential information | 10-97 | Check test signal |
| 10-36 | Correct time is ___ | 10-99 | Mission completed, all units secure |
| 10-37 | Wrecker needed at ___ | | |
| 10-38 | Ambulance needed at ___ | 10-100 | WC stop |
| 10-39 | Your message delivered | 10-200 | Police needed at ___ |
| 10-41 | Please tune to Channel ___ | 73's | Best wishes |
| 10-42 | Traffic accident at ___ | 88's | Love and kisses |
| 10-43 | Traffic tieup at ___ | | |

## British 934 MegaHertz CB Band

| Channels | | Channels | |
|---|---|---|---|
| 1 | 934.025 | 11 | 934.525 |
| 2 | 934.075 | 12 | 934.575 |
| 3 | 934.125 | 13 | 934.625 |
| 4 | 934.175 | 14 | 934.675 |
| 5 | 934.225 | 15 | 934.725 |
| 6 | 934.275 | 16 | 934.775 |
| 7 | 934.325 | 17 | 934.825 |
| 8 | 934.375 | 18 | 934.875 |
| 9 | 934.425 | 19 | 934.925 |
| 10 | 934.475 | 20 | 934.975 |

## British and American CB Channel Frequencies

| Ch. | British | American | Ch. | British | American |
|---|---|---|---|---|---|
| 1 | 27.60125 | 26.965 | 21 | 27.80125 | 27.215 |
| 2 | 27.61125 | 26.975 | 22 | 27.81125 | 27.225 |
| 3 | 27.62125 | 26.985 | 23 | 27.82125 | 27.255 |
| 4 | 27.63125 | 27.005 | 24 | 27.83125 | 27.235 |
| 5 | 27.64125 | 27.015 | 25 | 27.84125 | 27.245 |
| 6 | 27.65125 | 27.025 | 26 | 27.85125 | 27.265 |
| 7 | 27.66125 | 27.035 | 27 | 27.86125 | 27.275 |
| 8 | 27.67125 | 27.055 | 28 | 27.87125 | 27.285 |
| 9 | 27.68125 | 27.065 | 29 | 27.88125 | 27.295 |
| 10 | 27.69125 | 27.075 | 30 | 27.89125 | 27.305 |
| 11 | 27.70125 | 27.085 | 31 | 27.90125 | 27.315 |
| 12 | 27.71125 | 27.105 | 32 | 27.91125 | 27.325 |
| 13 | 27.72125 | 27.115 | 33 | 27.92125 | 27.335 |
| 14 | 27.73125 | 27.125 | 34 | 27.93125 | 27.345 |
| 15 | 27.74125 | 27.135 | 35 | 27.94125 | 27.355 |
| 16 | 27.75125 | 27.155 | 36 | 27.95125 | 27.365 |
| 17 | 27.76125 | 27.165 | 37 | 27.96125 | 27.375 |
| 18 | 27.77125 | 27.175 | 38 | 27.97125 | 27.385 |
| 19 | 27.78125 | 27.185 | 39 | 27.98125 | 27.395 |
| 20 | 27.79125 | 27.205 | 40 | 27.99125 | 27.405 |

# *Index*

Adjacent channel rejection 35, 36
AM 23, 24-25, 30, 33, 82-89, 90, 97
  98-99, 102, 105, 144-145, 146-147
Amateur Radio 24, 102-104
Amplifiers 97, 113
Antennas 29, 57-68, 44-53, 133-143
  base 57-68, 133-140
  beam 59, 61, 64-68, 137-143
  bilateral 113
  connection 125
  criss-cross beams 66
  do-it-yourself 133-143
  fibreglass 46
  gain 58
  gain vertical 134
  magnetic mount 50
  matchers 112
  mobile 44-53
  mount 48-50
  non sus 47-48
  quad 66
  stacked beams 65
  steel 45-46
  switchable beams 66
  switches 112
  twins 46-47
  UHF 48
  vertical beams 65
  whips 44-46
  ¼ wave 62, 133
  ⅝ wave 63
Anti-theft installations 42-43
Attenuators 25, 113, 35
Automatic modulation compressor 35
Automatic modulation limiter 35, 87
Base stations 54
BFO 88, 106
Bilateral amplifiers 113
Bleedover 35, 87, 117
Boosters 113
Breaking the channel 15
Burglar alarm 43
Burners 97
Capture effect 84
Carrier 75, 88
Channels 18-19, 34, 99, 147, 158
  hi/low 34
Channel jive 148-154
Channel-9 19, 34, 116
  priority channel-9 34
Channel selector 32
Clarifier 33, 88
Coax connectors 125-127
Coax and coax splices 51, 67, 125, 146

Connections
  antenna 125
  microphone 128
Continuity tester 129
Co-phase harness 47
Crystals 77, 80, 109
DC power supply 55
Delta Tune 33
Deviation 83, 111
Discriminator 75
Distant-Local Switch 33
Dual conversion IF 35
Dummy Loads 115
DX 25, 60, 90-104, 106, 107, 147
Effective radiated power 59, 113, 146
Electromagnetic spectrum 102
Eleven metre band 98
Emergency Procedures 116-117
Fading 95
Filters 35
Frequencies 18, 25, 90, 98, 102, 107, 109, 157
Frequency counters 113
FM 23, 30, 33, 72-77, 82-99, 105, 111, 144-145, 146-147
Gain 58-59, 63-66, 135, 137
Gooney birds 115
Ground plane 62
Ham license 24, 102-104
Handles 14
Heterodyning 73, 97
Home office 24-25, 27, 30, 35, 59, 98, 102-103, 108
IC chips 78
Installations
  anti-theft 42
  CB 40-43
  of base station 55-56
  quickie 50
  UHF CB 48
Intermediate frequency 73
License 24
Lightning 68
Limiter 75
Lingo 148
Loading coils 45
Long path 95
Meters
  modulation 111
  SWR 50
  field strength 111
  power 33, 111
  signal strength 33
Metric conversion chart 155

Microphone 34, 105, 124, 125, 128
  connections 125, 128
Mike gain 34
Mobile antenna 44-49
  as base 135
Mobile rig 38-43
  as a base 55-56
Mode switch 33
Modulation 13, 75
  AM 86, 75
  FM 82, 75
Morse code 103
Mount, antenna 48-50
  magnetic 50
  rain gutter 48
  locking 43
Negative earth installation 41
Noise limiter & blanker 32, 87
Noise suppression 130-132
Oscillator 73, 88
Overmodulation 35, 87
PA 34
PA horn 113
Phase lock loop 78
Phaser lasers 115
Phonetic alphabet 101
Pings 115
Polarisation 61, 147
Positive earth installation 42
Power microphones 87, 105, 124
Power reducers 25, 113
Power requirements 38
Preamplifiers 113
Procedures, emergency 116-117
Push up pole 62, 69
Q-signals 100-101
QSL cards 98-99
Radar detectors 113
Radio check 16
Receiving 32-34
Receiver preamps 113
Relay 76
RF gain knob 33
Roger bleeps 115
Rotators 70
R-S Reports 101
Rules and regulations 27
Scanners 107-109
Selectivity 35, 36

Sending 34-35
Sensitivity 35, 36, 106, 113
Short wave listening 90, 99, 104, 106
Single sideband 23-25, 30, 33, 82-90,
  98-103, 105, 106, 144-145, 146-147
Skip 23, 30, 31, 85, 91-104, 147
Skipland 23, 91
Slide mount 42-43
Sliders 98, 112
Soldering 125-127
Speakers, external 34, 112
Specifications 35-37
Speech compressors 105
Squelch 22, 32
Sunspots 90
  cycle 90-94
Suppression, vehicle noise 130-132
SWR 43, 49, 52-53, 110-111, 115,
  120, 143, 146
Ten calls 21
Ten code 156
Time conversion chart 155
Tone squelch 115
Tone switch 34
Towers & masts 168-170
Transceiver 13
Transistors 53, 97
Transmitter 13, 29
Troubleshooting
  guide 119-124
  receiving 119-122
  transmitting 123-124
TVI Filter 114-115
Twin Antennas 46-47
Type approved 24, 35
UHF-ultra high frequency 18, 24, 31,
  48, 60, 157
Valves 97
Vehicle noise suppression 130-132
VFO 98, 112
Volume off/on knob 32
Vox 89
Walkie-talkies 30, 31, 110-111
Warranty 38
Waves
  ground, sky, space 60
Whips 44-46
934 MHz UHF 18, 24, 31, 48, 60, 157

---

Educational and factual information on the state of the art is presented. It is up to the reader to determine the suitability of information for his or her intended use. It is not the intention of this book to encourage in any way actions which are contrary to existing or future regulations, rules, laws or local ordinances.

It's been a wild and crazy spin through CB Land. If you ever hear the Big Dummy on channel, give out a shout, and we'll modulate a while. Until then, we're putting the Golden Numbers on you, and we're on the side.